Where to Watch Mammals in Britain and Ireland

Richard Moores

Published 2007 by
A & C Black Publishers Ltd.,
38 Soho Square, London W1D 3HB

www.acblack.com

ISBN 978-0-7136-7161-2

A CIP catalogue record for this book is available from the British Library.

This book is produced using paper that is made from wood grown in managed sustainable forests. It is natural, renewable and recyclable. The logging and manufacturing processes conform to the environmental regulations of the country of origin.

Typeset by RefineCatch Limited, Bungay, Suffolk
Printed and bound in the United Kingdom

10 9 8 7 6 5 4 3 2 1

CONTENTS

ACKNOWLEDGEMENTS

The compilation of this guide inevitably involved considerable trawling of previously published information and I cannot thank enough the people who wrote the books, magazine articles, scientific papers, websites, county bird and mammal reports and reserves' information leaflets from which so much information was obtained. Particular mention must be made of David Macdonald and Fran Tattersall's recent publication *Britain's Mammals: The Challenge for Conservation* and their series of annual updates *The State of Britain's Mammals* that cannot be recommended highly enough to anyone with an interest in British mammals.

I am extremely grateful to Richard Webb for reviewing and improving the introductory chapters, species accounts and appendices, including updating the status and population information in line with Macdonald and Tattersall (2001) and Battersby (2005). I am also grateful to Bob Cornes for contributing extensively to the bat section including producing the sections on bat detectors and bat identification, and to Richard, Bob and Bill Underwood for providing details of numerous additional sites. Thanks also go to the following organisations and individuals for providing information on bat sites: Bat Conservation Ireland, Bedfordshire Bat Group, Birmingham and Black Country Bat Group, Dumfries and Galloway Bat Group, Isles of Scilly Wildlife Trust, Lincolnshire Bat Group, North Buckinghamshire Bat Group, Northamptonshire Bat Group, Northumberland Bat Group, Colin Pope (Isle of Wight), Phil Richardson and the South Lancashire Bat Group.

I would like to thank Nigel Redman, Marianne Taylor, Sharmila Logathas and Sophie Page at A & C Black Publishers for bringing this book to fruition, and to Ernest Garcia for the editing. I am also grateful to Brian Southern for his excellent maps, and for the re-use of maps from *Where to Watch Birds in Britain*, and the use of illustrations from the *Atlas of European Mammals*, published by T & A D Poyser. My thanks also to the artists James Gilroy and Dan Brown who have contributed illustrations. Last but not least I thank my parents for giving me a roof over my head during a time of need and for their encouragement during what was a long and often tiring project.

WHY WATCH MAMMALS?

Although many people find mammals exciting, few of us regularly go out specifically looking for mammals in Britain. For many, mammals are seen as a bonus when out looking for birds, plants or insects. This probably reflects the fact that Britain has a relatively small mammal fauna of around 70 species, and that most species are largely nocturnal and consequently difficult to see. Indeed, many zoologists have been happy to rely on signs, such as tracks, droppings, tree damage and discarded foodstuffs, to record the presence of mammals in an area, and many would argue that it is impossible to go out to look for most mammals. However, in reality many species can be seen with a little effort and patience.

Nevertheless, although many British mammals are difficult to see and lack the charisma of mammals elsewhere in the world they still deserve our attention. Watching wildlife and conservation go hand in hand. Many people gain an immense amount of pleasure from being able to go out and see a variety of species. Unfortunately, however, many species are rapidly declining in numbers and if we are to continue to be able to enjoy them in years to come we must help conserve the habitats on which they depend. Although conservation is heavily dependent on wildlife organisations carrying out major conservation projects these organisations still require the support of amateur naturalists and conservationists. Such support will either come from financial aid through memberships or, in many cases, from data from surveys or casual observations of animals and plants. The value of such information cannot be underestimated. There are relatively few professional conservationists and ecologists, and consequently the observations made by passionate amateurs are invaluable in assessing the distribution and status of many species. Recording mammals is as important in this respect as the bird surveys run by organisations such as the British Trust for Ornithology. I would encourage everyone to send their mammal records to the relevant county recorder and to participate in mammal surveys organised by The Mammal Society and Mammals Trust UK.

It is only through the gathering of such data that decision-makers can construct action plans for each species and the habitats on which they depend. Anyone can participate, and information provided by amateurs has led to some important discoveries. For example, a number of colonies of rare bats have been discovered in residential areas in recent years by enthusiasts and the general public. Such discoveries have increased our knowledge of bat populations and distribution and helped provide legal protection for these roosts.

Above all, many people simply watch mammals for the enjoyment that it gives them. Watching a Stoat or Weasel hunting along a hedgerow, or a family of Badgers emerging from a sett at dusk, gives many people immense pleasure. I hope that this book will help you gain experiences such as these on a regular basis.

FACTORS AFFECTING CONSERVATION:

Threats and the impact of introduced species

One of the main reasons for writing this book was to try to encourage people to go out to look specifically for mammals. Although many of the species in the British Isles are fairly common, particularly in a global context, others are declining and some may yet go the way of the Greater Mouse-eared Bat and cease to breed in this country. We consequently need to pay more attention to the plight of mammals in Britain and Ireland. I hope that this section will help to raise awareness of the threats faced by many species and encourage more people to become involved in mammal conservation.

A number of mammal species are under threat for a variety of reasons. Two of the most serious of these are direct persecution and habitat loss.

PERSECUTION

The history of British mammals is littered with examples of species' populations dwindling, many to extinction, at the hands of direct persecution. The Wolf, Brown Bear, Eurasian Lynx and Wild Boar were all once a part of the native fauna but were extirpated either because the human population perceived them as a threat or, in the case of the boar, through overhunting for food. Direct persecution continues to affect the populations of many British and Irish mammals detrimentally. The plight of the Wild Cat is a striking current example.

The (Scottish) Wild Cat is in serious trouble. This form of *Felis sylvestris* is currently competing with Iriomote Cat, Amur Leopard and Spanish Lynx for the title of most endangered felid in the world. Interbreeding with feral cats poses probably the most serious threat, but many populations are still being persecuted, with an estimated 10% of losses each year resulting from direct persecution. Other threats include snares, poison baits set for other species and losses to road traffic. Habitat change and the build-up of chemicals along the food chain have also been noted as potential threats to Wild Cat populations.

Other species are similarly affected by direct persecution. The Badger is hunted and 'controlled', illegally or otherwise, in most of the countries where it occurs. Indeed, Badgers face a plethora of threats including baiting, digging, lamping and shooting, snaring and trapping, poisoning and gassing. Badger populations have also been blamed (potentially totally unjustly) for spreading bovine tuberculosis amongst domestic cattle, and huge numbers have been culled in certain areas of Britain.

For a final example we should examine the role of gamekeepers. For centuries, gamekeepers have persecuted many mammal (as well as bird) species in order to protect their livelihood. Gamekeepers in upland areas control mammal species including Stoats, Weasels and Red Foxes: it has been estimated that perhaps 20,000 adult Red Foxes are killed annually in the Scottish hills. Such measures are seen as necessary by those carrying them out to protect their game species, be they partridges, pheasants, ducks or Red Grouse, for the shooting season and the revenue that it creates. It should be acknowledged, however, that gamekeepers do a huge amount of habitat management work, much of it benefiting a whole host of plant and animal species. It should also be remembered that if no one paid to shoot Red Grouse then the uplands would be a very different place indeed. The management of the heather moorland would cease. The patchwork of young and old heath created by annual burning would be replaced by ageing heather, where bracken and grasses would creep in followed by further habitat fragmentation. Populations of Red Grouse (an endemic race of the Willow Grouse) would dwindle, while internationally important populations of many upland waders would plummet due to habitat loss and a vast increase in predator numbers. It should be remembered that in Britain and Ireland such ecosystems are managed and controlled largely by humans, and predator control is often required in certain cases. Nevertheless, although sensitive gamekeeping is beneficial to wildlife interests, such activity must be conducted within the law, to avoid the persecution, intentional or otherwise, of protected species such as Wild Cats.

HABITAT LOSS

Habitat loss is one of the most serious causes of decline of a great many animal and plant species, and mammals are no exception. For example, although still occurring widely in England and Wales the Hazel Dormouse has declined, to the point of extinction in some areas, as a result of habitat destruction (although reintroduction schemes are underway in many areas). It is an animal of ancient deciduous woodland, dense shrubs and coppices offering a diverse range of foodstuffs. Arboreal pathways across woodland rides and tracks, and hedgerows linking woodlands, are crucial for the species to disperse successfully. Clear-felling, short-rotation coppicing and simple neglect have all contributed to the decline of Hazel Dormouse populations. Woodland management that leads to heavy shading and loss of the shrub layer has led to local extinctions, even when the wood itself survives. Fortunately many woods where Hazel Dormice still occur are now owned or managed by conservation organisations and are being managed in line with dormouse requirements. Unfortunately, however, many other woods are still being neglected due to the cessation of coppicing and, with the disappearance of hedgerows suitable for dispersal, this species still faces local extinction in many areas.

THE IMPACT OF INTRODUCED SPECIES ON NATIVE MAMMALS

A significant proportion of Britain and Ireland's mammal fauna is non-native, having been introduced from other parts of the world and establishing free-living populations. Introduced species include Fallow Deer, Sika Deer, Chinese Water Deer, Reeves's Muntjac, Red-necked Wallaby, Brown Hare, Rabbit, Grey Squirrel, Coypu, Black Rat, Brown Rat, Edible Dormouse, House Mouse and American Mink. These make up approximately 20% of the mammal fauna.

Some introductions have, however, been highly detrimental to the environment and/or the native fauna. Coypu, for example, were originally introduced into Britain for fur farming, but many escaped and a population became established in East Anglia. At its peak their population numbered 200,000 individuals, and they were prone to destroying crops and vegetation, as well as undermining banks and dykes. Their reign of destruction was brought to an end in 1989 when they were finally eradicated following a ten-year campaign by the Ministry of Agriculture, Fisheries and Food.

The damage caused by Grey Squirrels has also been well documented. Not only have they played a major role in the decline of Red Squirrels through direct competition and by spreading viruses, but they are also known to damage and destroy trees and crops and have an impact on songbirds by preying on eggs and nestlings.

More than half of our deer species have been introduced and most are now well established. Sika Deer were first introduced at the end of the nineteenth century and are now flourishing in many areas; they are also increasingly hybridising with native Red Deer. It is likely that pure populations of 'British' Red Deer will eventually only occur on Scottish offshore islands as a consequence of hybridisation of mainland populations with Sika and introduced 'Continental' Red Deer.

Deer have also been cited for causing environmental problems, and obviously these effects are magnified with larger populations. They have a tendency to eat trees and, more importantly, saplings, buds and other young growth, and so can alter the entire structure of a wood by reducing or preventing natural regeneration and by destroying the shrub layer and other plant species important for some birds and other woodland animals. The explosion in deer numbers has been cited as one of the main causes of declines in songbird populations within British woodlands.

American Mink escaped from fur farms and have become successfully established throughout Britain. They are thought to be one of the main factors behind the declines in Water Voles and the populations of some riparian bird species.

Rabbits have made themselves more or less irreplaceable. As a conservation tool Rabbits can be invaluable. They create and maintain short-turf habitats that would otherwise revert back to taller growth. These conditions are required by many of our rarer invertebrates particularly some of the downland butterflies. Many of the blues, particularly the Adonis Blue, the Chalkhill Blue and the reintroduced populations of the Large Blue, require short-turf conditions for their foodplants to grow and for their host-ant species to survive. Rabbits are also a major prey species for a range of predators, including such raptor species as Common Buzzards, Golden Eagles and Goshawks.

Last but not least, Wild Boar have recently re-established themselves, having escaped from farms, after an absence of some 700 years. Their impact is difficult to judge at present since the population is still quite small. Certainly they have the capacity to do considerable agricultural damage and the possibility of transmitting disease to domestic stock is a real one. Their rooting behaviour may have a serious effect upon woodland ground flora, with species such as the Bluebell and the Lady Orchid particularly at risk. On the plus side, this behaviour may help in the control of bracken, a species that has become a big enemy of the conservationist.

Most of these introduced species have been established for so long that we are used to their presence. Certain species have detrimental effects and it is important to recognise this. In such cases populations need to be controlled and in some cases this is being done, albeit at relatively low levels. What must be remembered is that populations of our native species should always have priority, and work to ensure their survival is extremely important.

THE SPECIAL PROBLEMS OF BAT CONSERVATION

The conservation of bats is one of our biggest challenges. Their nocturnal nature, and the difficulty of identifying many species, means that few of us ever see many species and the old expression 'out of sight, out of mind', is all too real for the 16 species of bat breeding in Britain and Ireland. Fortunately the Bat Conservation Trust and numerous local bat groups are dedicated to their conservation, and these organisations do a lot of good work for Chiroptera in this country. However, the number of people involved is still small and the legal protection afforded to bats carries little weight when lobbying against planning issues that may detrimentally affect bat species.

Threats to bats fall into three main categories. Habitat loss is the cause of the decline of many of our bat species. Landscape changes, particularly the loss of old pasture and deciduous woodland, and the improvement and clearance of watercourses, all affect foraging areas. We have already seen a huge reduction in linear features such as hedgerows. These provide vital corridors between roosts and foraging areas, and it is thought that many smaller bat species may be reluctant to cross large, open areas. The removal of hollow trees has a serious impact on tree-roosting species.

A second threat is from disturbance. If species roosting or hibernating underground are disturbed then their hibernation period is reduced, affecting their activity cycles. In addition, many roosts and hibernation sites have been lost as a result of work that has blocked off access to mines, caves, bridges and houses.

The final major threat to bats is from the use of pesticides. The decline in bat populations resulting from organochlorine pesticide poisoning has been a problem for the last few decades. Organochlorine residues are believed to be picked up from invertebrate prey, and small bat species are particularly at risk.

Nothing in conservation is ever straightforward and it must be realised that the British Isles have been modified by humans for centuries and will continue to be for the foreseeable future. Humans have the resources to control the populations of the majority of species at will. Such a concept is both exhilarating and scary. For example, the idea of transferable wildlife is exciting as we can now introduce or reintroduce many species to areas of suitable habitat or former strongholds and be reasonably confident that they will be successful. However, will such power mean we become complacent about wildlife? Will habitat management become lackadaisical? Will our new habitat-creation skills mean that some wildlife sites will be lost to industry or housing just because we can create new habitats and establish populations of what we want where we want? It is imperative these issues are acknowledged and addressed.

MAMMAL WATCHING AND THE LAW

The laws relating to British mammals are complex and cover many areas including handling, disturbing, shooting and hunting seasons, and trapping. Most legislation will have little impact on the mammal watcher but the following points should be remembered.

HAZEL DORMOUSE

Hazel Dormice are strictly protected by law and may not be intentionally killed, injured or disturbed in their nests except under licence. Consequently it is illegal to check nest boxes thought to contain Hazel Dormice, without a licence.

SHREWS

All shrews are protected under Schedule 6 of the Wildlife and Countryside Act (1981). Shrews may not be trapped without a licence and, when trapping for other small mammals, precautions must be taken to minimise the chances of killing or harming shrews. Information on trapping and the law is given in *Live Trapping Small Mammals: A practical guide* available from The Mammal Society.

PINE MARTEN

Pine Martens and their dens are fully protected by the Wildlife and Countryside Act (1981) and must not be disturbed other than under licence from Scottish Natural Heritage, the Countryside Council for Wales or Natural England (formerly English Nature).

WILD CAT

The Wildlife and Countryside Act (1981 and 1988) now gives strict legal protection to 'Scottish' Wild Cats and their dens. It is an offence to disturb a Wild Cat at the den. Unfortunately, difficulty in defining what a Wild Cat actually is (due to the problem of hybridisation with Feral Cats) means that in practice the law is rarely enforced. Illegal trapping, poisoning and shooting are all still major causes of deaths of Wild Cats.

BATS

In Great Britain, all bats are fully protected under Schedule 5 of the Wildlife and Countryside Act (1981). It is an offence for any person to disturb a bat roost intentionally or recklessly whether or not bats are present at the time. Bats may only be handled under licence.

GENERAL

Hunting with dogs is now outlawed in England, Scotland and Wales, and yet illegal hunting, particularly hare coursing, still occurs in many areas. Mammal watchers encountering illegal hunting, including badger baiting, should contact the local police. Contrary to popular belief many police forces will actively follow-up reports of illegal hunting.

FIELDCRAFT

Watching mammals is often frustrating. Other than the normally obliging Rabbits, Grey Squirrels and seals, obtaining good views of most mammals can be extremely difficult. When, for example, was the last time you saw a Weasel for more than a few seconds? Chance encounters, however brief, can be enjoyable, but what everyone really wants are prolonged views. The following suggestions should help you obtain longer observations.

TIMING

Generally speaking, the majority of mammals in Britain are nocturnal or crepuscular, i.e. most active at dawn and dusk, so this is obviously the best time to look. Having said this, many species will continue to be active at other times so searching should not be limited to early morning and late evening.

STEALTH

This is unquestionably the mammal watcher's most important attribute. The importance of patience and silent movement cannot be overstated. It must be remembered that although humans still depend quite heavily on hearing, it is our eyes that do a lot of the work and sight is the sense that we rely on most. For many mammals, the opposite is true. In closed habitats such as grassland (in the case of small mammals) and thick woodland, the ability to hear predators and potential prey is of great importance. It is very unlikely that you will see anything when making noise. Almost every species will hear you (and remember the majority have better hearing than we do) and simply take cover.

For many species smell is of equal importance to hearing. There is no point silently stalking a distant herd of deer on a hillside if the wind is behind you. Scent is a crucial part of predator detection, and they will soon be aware of you and be off. Always try to approach into the wind.

PATIENCE

Patience is a key prerequisite for the mammal watcher. For many species, just sitting quietly in a good area of habitat can reap rewards. Carnivores such as Stoats, Red Foxes and American Mink have well-used routes that they patrol each day looking for food. Stationing oneself in a spot overlooking such an area, e.g. a Rabbit warren, may enable you to obtain better views than you would normally expect. In some areas 'high seats' used by gamekeepers provide excellent vantage points and being raised reduces the risks of the animal detecting you. Please ensure that you obtain the permission of the gamekeeper or landowner before using such vantage points. Searching for mammals in areas of woodland can often prove a frustrating exercise. One rewarding technique is to sit quietly on a forest track or similar corridor that provides a gap in the tree cover. Mammals can often be seen crossing these gaps in order to patrol the full extent of their territories.

BAITING

Many mammals can be attracted by baiting an area with food, for example, raisins, seeds and cereals (including cornflakes) for rodents, apple for Water Voles, fly pupae for shrews, and strawberry jam for Pine Martens will often produce results. When out walking dropping a few seeds or fruits where a vole or mouse has just disappeared will often produce immediate results. Make sure that the spot that is baited can be viewed from an observation point (e.g. outside a window) and that the bait is topped up regularly. This is best done at dusk to reduce the risk of the food being taken by birds. Do ensure that the bait is placed either in or close to cover to minimise the risk of predation. If predators are a problem, placing a cover over the food, such as a mesh basket which still allows access to small mammals, is advisable.

SPOTLIGHTING

Given the nocturnal nature of many of Britain's mammals it is perhaps not surprising that spotlighting is an important aspect of mammal watching. Species such as Wild Cat, Pine Marten and Edible Dormouse are all generally easier to see at night. When spotlighting, the first indication of a mammal is generally eye-shine. However once you have spotted a mammal please remember not to shine the spotlight directly at the eyes and try to aim the beam just in front of the mammal to minimise disturbance. Wherever possible use green or red filters over the spotlight as mammals seem to be less disturbed by spotlights with filters. Finally, please remember it can be dangerous to spotlight along roads and please avoid spotlighting around houses or near to cattle, sheep, horses etc. **Under no circumstances should you use a spotlight in, or around the entrance to, a bat roost**.

CHECKING DEBRIS

Lifting up planks of wood, logs, corrugated iron sheets or other debris may reveal small mammals (and reptiles and amphibians). Unfortunately anything found in this way is unlikely to offer prolonged views, so if you want a good view and to secure a definite identification, the mammal should be caught. Ideally two people are needed to catch the mammal, one to lift the debris, and the other to grab the animal before it bolts. The best way of catching an animal is by using a net or an appropriate container. Alternatively the animal can be caught by hand. This, however, requires a considerable amount of speed and skill. In the interests of the animal's welfare you should pin it to the ground gently but firmly with the flat of one hand, grasping the scruff of the neck between finger and thumb of the other hand, then sliding the first hand back to grab the base of the tail.

BOTTLES

Small mammals frequently get trapped in discarded bottles and cans. You would have to be very lucky to find the mammal still alive (unless the trap has been deliberately set), so the animals will be in varying states of decay, but keys are available to identify the species from their remains.

TRAPPING SMALL MAMMALS

An excellent way to see small mammals is live trapping. There are a number of different types of trap on the market (see 'Equipment' for more details) and a useful booklet *Live Trapping Small Mammals: A practical guide* is available from The Mammal Society (details in 'Useful Addresses' section). Unfortunately, you need a large number of traps to achieve significant results, and the cost of the traps is likely to discourage all but the real enthusiast. It is, however, possible to trap animals occasionally with one or two traps. Alternatively, local mammal groups and organisations, such as The People's Trust for Endangered Species, Mammal Trust UK, and The Mammal Society, organise events to trap small mammals. It must be remembered that it is illegal to catch shrews without a licence, and care must be taken when catching other small mammals to minimise the chances of killing or maiming them.

It is also possible to use pitfall traps to survey the small mammals of a particular area. These are basically holes in the ground into which you hope an animal will fall. Pitfall traps should contain food and water and should be checked at least once every 24 hours.

LOOKING FOR SIGNS

There are many signs that can help the mammal watcher confirm the presence of a species in an area, and a number of books deal solely with this subject. The following are particularly recommended:

Finding and Identifying Mammals in Britain, G.B. Corbet (1989).
How to Find and Identify Mammals, G. Sargent & P. Morris (2003).
Mammals of Britain: Their tracks, trails and signs, M.J Lawrence & R.W. Brown (1974).

FOOTPRINTS

Mammal tracks can vary considerably depending on how the animal was moving and the nature of the ground. Although even experts cannot reliably identify every species, some tracks can be identified. Locating well-worn tracks can help identify vantage points from which to look for the mammals themselves.

NESTS

Although many nests can only be identified in the presence of their owners others, for example, squirrel dreys and the nests of Harvest Mice and Hazel Dormice, are readily identifiable with experience. Harvest Mouse nests are most easily located in winter in hedgerows and fences, among long grass or in reedbeds. They are very small, spherical balls of woven grass, about 5–10cm in diameter and usually about 10–50cm above the ground. Hazel Dormouse nests are less easily found but may be located in the branches of small trees and shrubs, especially where there is a dense tangle of creepers in close proximity. Hazel Dormice will also make use of old birds' nests and nest boxes (the latter are also occupied by mice, voles and occasionally Weasels in winter). It must be remembered that the inspection of Hazel Dormouse boxes is only permitted under licence, although organisations such as The People's Trust for Endangered Species, Mammals Trust UK and The Mammal Society organise events to see dormice. The Mammal Society also organises courses for individuals wanting to obtain dormouse licences. Please remember that mice can give you a nasty bite so gloves are recommended when checking boxes in winter.

DROPPINGS

Droppings are a good way of determining the presence of species in an area. Otter spraints, for example, are fairly distinctive: they are black when fresh but rapidly bleach, and are placed in prominent points such as boulders in a stream or on grass tussocks. Field Vole droppings can often be found along runways: they are about 5mm long with rounded edges. Bat roosts can often be detected by the accumulation of droppings.

BURROWS

With experience many mammal burrows can be identified to species, for example, Badger setts are easy to identify. They consist of a number of burrows in close proximity to one another, each with a large mound of earth in front of the hole. There are usually other signs such as tracks, droppings or food remains to aid the identification of a burrow's owner.

FOOD REMAINS

Several species can be reliably identified from food remains, e.g. by checking hazelnuts for distinctive marks. These include the Hazel Dormouse, Wood Mouse, Bank Vole and Grey Squirrel.

BIRD PELLETS

Remains of small rodents can often be found in the pellets deposited by owls and raptors. Such remains usually consist of bones and fur and identification can be achieved by isolating any skulls (if still intact) or jawbones.

ROAD CASUALTIES

Searching roads for mammals that have been the victims of motor vehicles can confirm the presence of a species in an area. It has been particularly useful in documenting the recolonisation of England by Polecats.

WATCHING BATS

Watching and identifying bats is extremely difficult. Searching suitable habitat at dusk will usually reveal some foraging species, although identifying bats in flight with any certainty is extremely difficult without enormous experience. A good way to watch bats is to find potential roost sites and wait for them to emerge. Bridges, caves, hollow trees and buildings can all harbour bat roosts and all are therefore worth watching. Under no circumstances should anyone enter a known roost site, unless accompanied by a licensed bat worker. To do so risks disturbing the bats, which is in any case illegal. Anyone visiting any of the roosts described in the site guide should only watch bats emerging.

Without doubt a bat detector is the most valuable piece of equipment for studying bats. Bats emit echolocation calls in order to locate prey and these sounds can be picked up on bat detectors. Details of the frequencies on which bats can be heard can be found in the section on bats. There are a number of different types of bat detector and it is worth going out with a local bat group to see the different types in action before buying one yourself. Alana Ecology's website, www.alanaecology.com, provides an excellent overview of the different types available.

The netting, handling and marking of bats can only be done with a licence. Please refer to your local bat group for details. The best way of seeing and learning about bats is to join your local bat group on field trips. Alternatively The Mammal Society, Mammals Trust UK (MTUK) and the Peoples' Trust for Endangered Species (PTES) sometimes organise events to go out looking for bats. Please refer to their websites for more details.

SAFETY WHEN BAT WATCHING

Good bat sites are often hazardous places, not least because they must be visited in the dark. It is always advisable to familiarise yourself with the site in daylight before attempting to watch bats. With a small site, that may simply be a matter of arriving half an hour before sunset. Bats may sometimes emerge surprisingly early, especially in midsummer, and the only chance of seeing bats clearly is when there is still good light in the sky. With larger, more complex and more hazardous sites, it may be advisable to make a preliminary visit in good daylight before attempting any bat watching. This is especially true for sites that have open water and for those in dense woodland. Dawn surveys should always be preceded by a daylight visit, because of the extra difficulty of arriving in darkness, and having to find your way.

Crime and vandalism may be factors to consider in more urban sites such as parks. Floodlit areas may be safer, but they are likely to have limited bat activity in many cases. The overwhelming advice is not to bat watch alone. All the hazards are more manageable when you are with others, and you will also probably see more bats. This is a little different from the situation when watching terrestrial mammals, when more people means more disturbance. As long as you avoid excessive use of torches and do not watch too close to roosts, you are unlikely to disturb bats.

EQUIPMENT

Finding and watching mammals can take many forms, and different species require different techniques. In addition to attributes such as persistence, stealth or straightforward good fortune there are a number of items that may assist or enhance the experience of finding a species.

Binoculars are essential for watching mammals. Given that many species are either crepuscular or nocturnal, binoculars with good light-gathering capabilities (e.g. 7 × 42) are perhaps more important than those with high magnification.

Telescopes can be useful in a number of circumstances where a species is only viewable at long range, for example, for obtaining better views of a distant herd of deer, for scanning mountain ridges in search of Mountain Hares or for observing cetaceans and seals offshore. Wide-angle lenses of around 30× magnification are probably the best choices but a lower magnification eyepiece with a wide objective lens may give better light gathering in poor light.

Cameras are an underestimated piece of equipment. The main purpose of carrying a camera in the field is to get shots of species whose identity may not be immediately obvious. A camera can be particularly useful when trapping small mammals and for camera trapping: i.e. placing a camera in a hidden position where it can remotely take pictures of difficult-to-see animals. This is mainly useful in surveying/monitoring of species. Camera trapping is, however, expensive and should not be undertaken lightly.

Longworth traps have traditionally been used for small mammal trapping. Unfortunately, given the need for large numbers of traps to cover an area, their high cost puts them out of reach of most casual observers. Fortunately, cheaper traps such as the Field Trip Trap and the Sherman are now available although even these will require a significant financial outlay. For details of a range of traps and the current prices visit Alana Ecology's website at www.alanaecology.com. More information on small mammal trapping can be found in the 'Fieldcraft' section.

Bat detectors are highly specialised pieces of equipment that can be used to identify bats by converting their ultrasonic foraging calls into audible sounds. They can also be used to detect shrews, which is particularly useful for people who have lost the ability to pick up high frequency sounds. There are various types available and they can be obtained through ecological equipment suppliers including Alana Ecology who stock a range of detectors. Their website provides useful information on the different types available. See pages 31–37.

Nets can be useful when looking for small mammals under logs, corrugated iron sheets, or other debris. They can be used to capture any small mammal that is located for closer inspection.

Plastic bags come in handy when collecting mammals signs for future inspection. Specialist bags can also be used to hold mammals that have been caught in traps or found under logs or corrugated iron sheets.

Magnifying glasses are particularly useful when studying mammal signs. Droppings, pellets, hairs and tracks can all be viewed in better detail with a magnifying glass.

Night vision equipment Night vision equipment, although often expensive, can be very useful when badger-watching or looking for other mammals. There are several generations of night vision equipment with prices ranging from as little as £180 up to £3,000 or more. Night vision equipment can take a lot of getting used to but it is well worth the effort. For details of a variety of night vision equipment try www.alanaecology.com, www.warehouseexpress.com or www.sovietbazaar.co.uk.

Spotlights High-powered spotlights are invaluable for watching nocturnal mammals, especially when used in conjunction with green or red filters. Before spotlighting over private land permission should be sought. Spotlights with 1,000,000+ candlepower are ideal. If you are able to spotlight from a vehicle, e.g. along forest trails, spotlights running off car batteries are the best option as they will last longer than stand-alone spotlights. If spotlighting on foot, head torches are ideal for picking up eye-shine. You can then switch to more powerful handheld spotlights once you have located a mammal. It is best not to aim the beam directly at the animal but should be aimed just off of it to minimise disturbance to the animal. Powerful spotlights should never be shone at bats, and lights of any kind (even small torches) should not be shone at bat roosts. Species such as Edible Dormice are more likely to give prolonged views if a filter is used. In addition try to avoid spotlighting near houses for obvious reasons. Unfortunately, few manufacturers produce high-powered rechargeable spotlights with sufficient battery life so it may be better simply to use top of the range Maglite flashlights or spotlights available from motoring suppliers etc. For higher quality spotlights try www.alanaecology.com or search the internet for the Australian Lightforce range of spotlights which are highly recommended. They are widely available in the UK.

INTRODUCTION TO THE SPECIES ACCOUNTS

Each group account includes:

- A list of the British species, followed by an introduction to the group.
- An overview of how to separate similar species.
- An overview of techniques useful in seeing the species.
- Key sites for seeing the species.
- A table showing the latest population estimates, the current range of each species and its habitat preferences.

The bat account includes additional sections on bat detectors and the sounds that bats make. For the sake of completeness, the sections on bats, seals and cetaceans include details of vagrant species.

Where appropriate, species have been grouped together, e.g. 'bats', 'shrews', 'rodents' etc., as many of the techniques for seeing a particular species, such as trapping for voles and mice or using bat detectors, are common to an entire group. Watching mammals is often more a question of luck than tactics, but the suggestions provided should increase your chances of success.

The information draws heavily on *Britain's Mammals: The Challenge for Conservation* (Macdonald & Tattersall, 2001), and I strongly recommend that anyone wanting more information consults this book and the subsequent annual updates (Macdonald & Tattersall, 2001–2004, Macdonald & Baker, 2005). Population estimates are also taken from *UK Mammals: Species Status and Population Trends* (Battersby, 2005).

The order of mammals in this book follows the sequence used in *Mammals of the World: A Checklist* (Duff & Lawson, 2004). The species names used in the introductory sections also follow Duff & Lawson (2004), but thereafter many of the (bracketed) prefixes are dropped to reflect common usage in the British Isles.

INSECTIVORES

(Western) Hedgehog *Erinaceus europaeus*
(European) Mole *Talpa europaea*
(Eurasian) Water Shrew *Neomys fodiens*
Common Shrew *Sorex araneus*
(Eurasian) Pygmy Shrew *Sorex minutus*
Lesser White-toothed Shrew *Sorex coronatus*

Six species of terrestrial insectivore (excluding bats) occur in the British Isles but, with the exception of the Hedgehog, most are small and can be difficult to see well.

Hedgehogs are extremely widespread and should be encountered without too many problems although the 2005 survey by the Mammals Trust UK, Living with Mammals, has revealed a huge decline in the east and north-west of England, with significant declines in many other areas. Hedgehogs on the Outer Hebrides have become a conservation issue as they do considerable damage through the predation of nests of ground-nesting birds, and culls are underway in some areas there.

Moles, though widespread and common, are seldom seen due to their subterranean lifestyle, although their molehills are ubiquitous and sometimes a nuisance to gardeners.

All four shrews can be seen with a little effort although obtaining prolonged views can be difficult.

IDENTIFICATION

Hedgehog Totally distinctive and unlikely to be confused with any other species.

Mole Totally distinctive and unlikely to be confused with any other species.

Water Shrew The largest British shrew with black dorsal fur, silvery-grey underparts with a sharp demarcation in adults. Can show a white patch above the eye.

Common Shrew Larger than Pygmy Shrew with proportionately shorter tail. Common Shrew normally shows a sharp contrast between the back and the flanks. Can show a distinct band of intermediate colour between the dark upperparts and pale underparts.

Pygmy Shrew Normally paler and greyer than Common Shrew and lacks the intermediate band between the upperparts and underparts. Has a proportionately longer and thick er tail than Common Shrew, about 65% of the head and body length. The head is more bulbous than that of Common Shrew with a relatively short, narrow snout.

Lesser White-toothed Shrew The only shrew on the Isles of Scilly.

SEEING INSECTIVORES

Hedgehogs are almost entirely nocturnal, but may occasionally be seen at dawn and dusk. Unfortunately most encounters tend to be with flattened individuals at the sides of roads. They usually hibernate during the winter months from December until March/April although during mild spells it is possible to discover active Hedgehogs in any month of the year. They survive well in many gardens, particularly where food, e.g. cat food (recommended) and milk, is put out for them and this is probably one of the best ways of observing these creatures. When moving through low vegetation and over dead leaves they produce a characteristic shuffling noise that is easy to recognise once you are familiar with it.

Moles are extremely difficult to see because of their subterranean lifestyle, despite being common. The presence of molehills indicates the presence of Moles in an area, and they can occasionally be seen as they excavate these hills, particularly after rain. They do also occasionally come to the surface, e.g. during juvenile dispersal, and to collect nesting material – dry grass and leaves – and also to look for earthworms when the soil is dry. May and June are probably the best months to see Moles on the surface. It is also possible to go live trapping with local pest-control officers. While most pest-control officers trap and kill Moles, in some areas they are live-trapped and re-released away from the problem area.

Water Shrews can be seen throughout the day and sitting quietly in suitable habitat may produce sightings. Family parties are especially visible from late April to early June particularly in the evening.

Shrews are active by day and by night but tend to be more active from dusk to dawn. In winter some spend as much as 80% of the time underground. All the shrews utter high-pitched aggressive cries which are often the first indication that there is a shrew nearby. Unfortunately as we grow older we often lose our ability to pick up high-frequency calls. This can lead to a false perception that shrews are actually becoming scarcer in an area. Fortunately bat detectors can help overcome this as they are just as effective for picking up shrew calls. Shrews can also regularly be found by turning over logs and planks found lying in suitable areas. Small mammal trapping provides another opportunity to see shrews but please remember that shrews can only be trapped under licence. If you do have the appropriate licence it is important to bait with the correct food, e.g. maggots or casters, as shrews are very fragile and need to eat their own body weight of food each day. See Fieldcraft and Equipment for further details.

Lesser White-toothed Shrews are mainly nocturnal but tend to be more diurnal in summer. They are extremely common on the Isles of Scilly and with a bit of effort may be seen quite easily. One of the best ways to look for them is to check beaches such as that at Porthloo on St Mary's, carefully turning over some of the larger boulders and rocks at the top of the beach. Views may be brief but shrews shelter under these rocks. Another method is to walk around slowly and quietly at the top of the beach and to listen for the shrill high-pitched calls of the shrews. Once you locate a shrew by call sit nearby and watch for shrews running between large boulders normally at the top of the beach. Prolonged views are more likely using this method. The sandhoppers that occur on such sandy beaches are a favoured food item.

KEY SITES

Most species of insectivore are widely distributed and can be seen in many localities.

Water Shrews are less common but are regularly encountered at the following sites:

South-west England
Five Acres CWT Reserve, Cornwall

South-east England
Elmley Marshes RSPB Reserve, Kent
Stodmarsh NNR, Kent

East Anglia
Fowlmere RSPB Reserve, Cambridgeshire
Wicken Fen NT, Cambridgeshire

Midlands
Barnwell Country Park, Northamptonshire
Cotswold Water Park, Gloucestershire/Wiltshire
Rye Meads RSPB Reserve, Hertfordshire
Saltfleetby-Theddlethorpe Dunes NNR, Lincolnshire

Northern England
Blacktoft Sands RSPB Reserve, East Yorkshire
Leighton Moss RSPB Reserve, Lancashire
Parkgate Marsh, Cheshire

Wales
Dan-y-Graig NR, Gwent
Teifi Marshes NR, Ceredigion
Tregaron Bog/Cors Caron NNR, Ceredigion

Scotland
Ardnamurchan Peninsula, Argyll and Bute

STATUS AND DISTRIBUTION

Hedgehog
Population 1,560,000*
Range Common throughout mainland Britain and Ireland. Introduced to many islands including Orkney, Shetland, Isle of Man and some of the Channel Islands.
Habitat Parks, gardens, farmland, woodland edges, hedgerows and suburban habitats where there is plenty of food.

Mole
Population 31,000,000
Range Throughout mainland Britain. Absent from Ireland and most islands except Anglesey, the Isle of Wight and some of the Inner Hebrides.
Habitat Pasture, arable farmland and deciduous woodland up to 1000m. Less common in coniferous forests, moorland and sand dunes.

Water Shrew
Population 1,243,000
Range Throughout mainland Britain although local in northern Scotland. Also found on the Isle of Wight, Anglesey, Skye, Mull and Arran.
Habitat Wetlands including rivers, ponds, drainage ditches, rocky shorelines, reedbeds and fen. Fast-flowing waters seem to be preferred in some areas. Studies have shown them to be more common away from wetland habitats than previously realised.

Common Shrew
Population 41,700,000
Range Widespread on the mainland. Absent from the Isles of Scilly, Orkney, the Outer Hebrides, Shetland and parts of the Inner Hebrides.
Habitat Hedgerows, fields, woods, heath, dunes and other scrubby areas but fairly scarce on moors. Prefer moist conditions.

Pygmy Shrew
Population 8,600,000*
Range Found throughout Britain including the Orkneys and the Outer Hebrides although absent from the Isles of Scilly and Shetland. The only shrew found in Ireland and on Lundy.
Habitat Found widely where there is adequate ground cover, e.g. heath, grassland, sand dunes and woodland edge although less common in woodland than Common Shrews. Commoner in grasslands than Common Shrews.

Lesser White-toothed Shrew
Population 40,000
Range Restricted to the Isles of Scilly and Jersey and Sark in the Channel Islands
Habitat Wide range of habitats including grassland, heath, coniferous plantations, sand dunes and sandy beaches.

*Excluding Ireland

BATS

Greater Horseshoe Bat *Rhinolophus ferrumequinum*
Lesser Horseshoe Bat *Rhinilophus hipposideros*
(European) Whiskered Bat *Myotis mystacinus*
Brandt's Bat *Myotis brandtii*
Natterer's Bat *Myotis nattereri*
Bechstein's Bat *Myotis bechsteini*
Daubenton's Bat *Myotis daubentoni*
Greater Mouse-eared Bat *Myotis myotis*
(Common) Serotine *Eptesicus serotinus*
(Common) Noctule *Nyctalus noctula*
Leisler's Bat *Nyctalus leisleri*
Common Pipistrelle *Pipistrellus pipistrellus*
Soprano Pipistrelle *Pipistrellus pygmaeus*
Nathusius's Pipistrelle *Pipistrellus nathusii*
(Western) Barbastelle *Barbastella barbastellus*
Brown Long-eared Bat *Plecotus auritus*
Grey Long-eared Bat *Plecotus austriacus*

Bats are without doubt the most challenging group of mammals to see and identify in Britain and Ireland. They are frequently seen but are difficult to identify in the field, even with specialist equipment. On the other hand, if you have a bat detector and a little knowledge, they can be very rewarding, with the thrill of making genuinely new observations a real possibility.

There are currently 16 species breeding in Britain and Ireland, and one recently extinct species. Although one or two are readily recognisable in the field – most notably Noctules with their characteristic fast flight action, and Daubenton's Bats with their hovercraft-like flight low over the water – most others can only be safely identified in the hand or with the aid of a bat detector. In many cases the use of bat detectors greatly adds to the experience, with species such as Greater Horseshoe Bat having particularly wacky echolocation calls. They also considerably extend the encounters because the bats can often be heard when it would be difficult to see them without the clue provided by the detector.

Consequently the best way to see and, more importantly, to identify bats is to join a local bat group or other organisation on a bat walk. This will often involve going out with experienced bat workers and using bat detectors to identify species feeding in a particular area, or checking bat boxes or winter hibernation sites such as caves and ice houses for roosting animals, in which case you will normally be able to observe a range of species close up. Details of local bat groups can be found on the website of the Bat Conservation Trust: www.bats.org.uk. Many of these offer walks during the summer months. Organisations such as the People's Trust for Endangered Species, Mammal Trust UK and The Mammal Society also offer walks to see bats, including some of the rarer species such as Bechstein's. See Useful Addresses for more details.

For those who want to go it alone a bat detector is essential. A wide variety are available, and again it is often worth joining a local bat group on one of their walks, to see the different types in action and to learn how to use them, before spending any money on one yourself.

IDENTIFICATION POINTERS

Species	Size/g	Roost types	Emerge	Foraging habitat	Feeding strategy*
Greater Horseshoe Bat	28	Buildings, caves	Late	Woodland, edges	Hawking, pouncing
Lesser Horseshoe Bat	7	Buildings, caves	Late	Woodland, edges	Hawking, pouncing
Whiskered Bat	6	Buildings, trees	Fairly late	Woodland, edges, riparian	Hawking
Brandt's Bat	7	Buildings, trees	Fairly late	Woodland, edges	Hawking
Natterer's Bat	9	Trees, buildings	Late	Woodland, over water	Hawking, gleaning
Bechstein's Bat	10	Trees	Fairly late	Woodland	Gleaning, hawking
Daubenton's Bat	8	Trees, bridges	Late	Over water, woodland	Gaffing, hawking
Serotine	25	Buildings	Early	Open – pasture & parkland, edges	Hawking, pouncing
Noctule	30	Trees, buildings	Very early	Open, over water	Stooping, hawking
Leisler's Bat	15	Buildings, trees	Very early	Open	Stooping, hawking
Common Pipistrelle	5	Buildings	Early	General, edges	Hawking
Soprano Pipistrelle	5	Buildings	Early	Woodland, edges, riparian, general	Hawking
Nathusius's Pipistrelle	7	Trees, buildings	Early	Woodland, over water	Hawking
Barbastelle	9	Trees in woodland	Early	Woodland, open, edges, riparian	Hawking
Brown Long-eared Bat	9	Buildings, trees	Late	Woodland	Gleaning, pouncing, hawking
Grey Long-eared Bat	10	Buildings, trees	Late	Woodland	Gleaning, pouncing, hawking

*Feeding strategies defined:
Gaffing – trawling for insects from the surface of water by gaffing them with large feet or the tail membrane.
Hawking – taking insects directly out of the air.
Gleaning – a comparatively uncommon style of foraging in which bats detect, locate and capture their prey on surfaces (usually of trees).
Pouncing – prey caught when on the ground.
Stooping – prey caught in the air following a stoop.

Handling wild bats is an offence. The only exceptions are to release them from the living space of a house, or to rescue injured animals. In those circumstances, it is advisable to wear gloves and avoid direct contact with the bat since there is a possibility of disease transmission. It must be remembered that disturbing bats at either summer or winter roosts is also an offence carrying significant financial penalties. If visiting such sites you must be accompanied by a licensed bat worker.

SEEING BATS

Bat roosts (both summer and winter) are protected by law from disturbance, for the very good reason that it may cause the bats to leave the roost, with possible disruption to breeding or other activities. Bats are very slow to reproduce, with females producing only one young per year at most, and such disruption can have a catastrophic effect on the numbers of bats in the colony, especially if repeated. Entering a bat roost is therefore not feasible, except in special circumstances when accompanied by a bat licence holder. Shining lights at the roost exit and making loud noises, especially at higher frequencies, must also be avoided. Such disturbance is not only illegal but will also probably have the effect of delaying the departure of the bats, so that it will be too dark to see them when they do emerge. However, careful observation outside roosts can be an effective way of seeing bats and should not cause a problem if done sensibly.

Foraging areas Good foraging areas can attract large numbers of bats, although individuals may not stay for long before moving off to other areas. Different species have different preferences, of course, but there are plenty of places that are good for a range of species.

1. *Pasture* – particularly favoured by some specialist feeders (**Serotines** and **Greater Horseshoe Bats**)
2. *Meadows and grassy areas* – especially if surrounded by hedges or taller vegetation (**Common Pipistrelles, Noctules, Leisler's Bats**)
3. *Gardens* – especially if they contain a mixture of shrubs and more open areas (**Common Pipistrelles**)
4. *Ponds and lakes* – especially if bordered by trees (**Daubenton's Bats** low over the water, **Natterer's Bats** and **Common & Soprano Pipistrelles** higher, **Noctules** very high)
5. *Streams and rivers* – preferably without turbulent water flow, and especially if lined with trees or tall vegetation (**Daubenton's Bats**)
6. *Woodlands* – difficult to watch in but have some of the rarer and more specialist bats (**Brown Long-eared Bats, Natterer's Bats, Brandt's Bats, Barbastelles, Bechstein's Bats, Lesser Horseshoe Bats**)
7. *Edges and hedges* (**Common Pipistrelles, Whiskered Bats**)

Commuting routes Bats use linear features of the landscape as commuting routes to move between roosting areas and foraging areas. They can therefore be productive for bat watching, especially shortly after sunset or before dawn. The bats will rarely stay for long, frequently flying rapidly in a straight line. Be prepared for some bats to fly surprisingly low (sometimes as low as knee height) along hedgerows or woodland rides. Commuting routes may include woodland paths and rides, hedgerows (especially if tall or broad), tree lines, streams and rivers, ditches and dykes, and lanes, especially if sunken or enclosed by tall vegetation.

Patterns of bat activity

March – April	Leave hibernation and begin to feed.
May – June	Females gather in maternity roosts.
June – July	Females give birth. Young start to fly about 3 weeks later.
August – September	Feed intensively. Swarm around hibernation sites.
October – November	Prepare for hibernation.
December – February/ March	Hibernation.

THE BAT WATCHER'S YEAR

Spring and summer

From the middle of spring onwards, bats are active and can be seen in their feeding areas. This is the time to get out the bat detector and start looking for bat activity. The first few evenings are best spent at sites that are good for the more common species. There are few of us that do not need a period of refamiliarisation each spring with the sounds from a bat detector, and the best way to get it is to go to sites that you know. After that, it is time to start looking in more difficult sites, such as woodland, and for the less common species.

In May and June, female bats of most species will begin to gather in maternity roosts. Observation of emergence will give you good views and an opportunity to practice your bat detector skills. It can also be very revealing to do emergence counts, and this information can contribute to the National Bat Monitoring Programme. See the Bat Conservation Trust website for details of how to take part. This is also a good time of year for finding roosts, by looking for swarming bats around a roost at dawn.

During June and July, a changing pattern of activity can be seen as the females give birth. They frequently return to the roost, extending the opportunity to watch and listen. It is especially important to avoid disturbance during this time. In particular, the bat watcher must avoid shining torches or making loud noises near roosts. With care, you can see a great deal at this time. It is always interesting to see how quickly lactating females return to the roost after emerging and feeding. When the young bats first begin to fly, about three to four weeks after birth, the numbers of bats leaving the roost at dusk suddenly increases. The young bats follow their mothers and it is frequently possible to see bats in pairs as they fly away from the roost area.

Later summer and autumn

Feeding activity continues long into the autumn but new patterns of behaviour occur in August and September. This is the peak mating time and male pipistrelles can be seen 'songflighting'. They fly up and down a regular beat, perhaps a woodland ride or a path beside a pond. With a bat detector tuned down to 30 kHz, you can hear the social calls as they advertise their presence to the females. Sometimes the males hang up in a tree and continue to produce loud social calls. Another recently discovered pattern of behaviour at this time of year is autumnal swarming. Bats of many species swarm, sometimes in large numbers, outside the entrance to hibernation sites. This can be hard to see, because it tends to occur very late into the night, but it can be heard on a bat detector. The function of the behaviour is uncertain but it may be related to courtship behaviour and mating.

As autumn continues, the bats' behaviour is increasingly weather-related. On warm evenings, bats will be very active, as they feed rapidly to build up their weight for the period of hibernation. This can give scope for good bat watching, but the early dusk and unpredictability of conditions make for difficulties. On cold autumn evenings, there may be little or no bat activity.

Winter

Bats enter hibernation around November. They hibernate in caves and other underground sites, in buildings, or in trees, depending on the species. Hibernating bats revive naturally from time to time, and may leave the hibernation site and fly around to feed if food is available. The occasional flying bat can be seen at any stage of the winter, even sometimes in daylight. However, their limited food reserves allow for only a very few revivals during one winter, because each requires a large amount of energy to raise the body temperature. Entering a hibernation site or disturbing bats in hibernation is against the law except for very small groups accompanying a Natural England licence-holder. Disturbing hibernating bats can lead to their death, by causing them to use up excessive food reserves as they warm up in response to the disturbance. Hibernation lasts until around March, depending on weather conditions. When in hibernation, they are therefore unwatchable, except during the carefully controlled hibernation surveys led by licensed bat workers. If you join a local bat group, you may be able to take part in one of these surveys. Winter is generally a time of inactivity for bat watchers, as well as for bats. Switch your attention to other mammals that are easier to see in winter than summer.

EQUIPMENT FOR BAT WATCHING

The one major item that is indispensable for serious bat watching is a bat detector. A good torch is important for your own safety if you are out in the countryside after dark but it should not be shone at flying bats or at bat roosts. The disturbance caused is not only harmful to the bats but will also prevent you from seeing very much since they will avoid the light. Contrary to myth, bats have good eyesight: probably about as good as human vision in conditions of limited light, although they cannot see in colour. Perhaps surprisingly, binoculars, night vision equipment and infrared sensitive video have very limited use. They can be helpful when you know exactly where the bats will be, such as flying over a pond, but even then they can be very difficult to use due to the speed of bats' flight, focusing difficulties and the effects of residual light in the sky. Photography of roosting bats is an offence, unless done under licence from Natural England.

BAT DETECTORS

A bat detector is very useful for finding bats and essential for starting to identify them in flight. All bat detectors convert the bats' ultrasonic calls into an audible sound that can be listened to through a speaker or headphones, or recorded to tape, minidisc or a digital recording device. By using a bat detector, you can learn to recognise the calls that bats make and to distinguish at least some of the species. Recordings can be used to identify bats afterwards, especially if they are made from the more expensive frequency division or time expansion bat detectors, which allow sonograms to be produced after downloading to a computer with appropriate software installed.

TYPES OF BAT DETECTOR

1. Heterodyne. A heterodyne detector generates an electronic tone that it subtracts from the detected sound, leaving an audible frequency that conveys some characteristics of the sound. It effectively detects a narrow slice of the call ranging about 5kHz above and below the generated frequency. The frequency of the generated tone can be tuned in order to listen for different call frequencies.

2. Frequency division. A frequency division detector electronically reduces all frequencies in the detected sound by a factor of 10, giving an audible sound that contains a full range of frequencies but has lost some detail.

3. Time expansion. A time expansion detector performs the electronic equivalent of slowing down a tape recorder so that all frequencies in the detected sound are divided by a constant factor (x10, x20 or x32), with the duration of the call being increased by the same factor. This converts the call to an audible frequency while retaining all key aspects of the sound.

COMPARISON OF THE THREE TYPES OF BAT DETECTOR

	Heterodyne	Frequency division	Time expansion
Real time (continuous listening)	Yes	Yes	No
Frequency range	Narrowband (tunable)	Broadband	Broadband
All species simultaneously	No	Yes	Yes
Background noise level	Low–Moderate	High	Low
Enables identifaction by ear	Yes	To some extent	Yes
Capable of producing sonograms	No	Low quality	High quality
Cost	Moderate (less than £150)	Reasonably moderate (less than £250)	High (more than £500)

Some bat detectors can operate in two or three different modes, giving for example heterodyne and frequency division information simultaneously. The latest development is detectors with built-in digital memory, making it very easy to record what you hear. Older detectors require an additional recording device if you wish to make recordings. Before buying a detector, it is worth perusing the relevant websites to find out what is available, whether new models are imminent, and of course prices. A local bat group can probably give you both advice and a chance to try out a detector before buying.

SLAPS, CLICKS, TICKS, TOCKS AND WARBLES – LEARNING TO RECOGNISE DIFFERENT CALLS ON A BAT DETECTOR

Bat detectors are wonderful gadgets. They make the impossible possible: they allow us to hear the ultrasound generated by bats, revealing the fact that, far from being silent, most species are remarkably noisy creatures. In most cases, bat detectors live up to their name, in that they detect bats very effectively. You will see far more bats if you are using a detector, and you will hear others that you are unable to see, particularly in woodland. If you want to do more than simply enjoy watching an unknown species of bat flying on a summer's evening you will need to use a bat detector, but you will also need to acquire a significant amount of experience before you can make any reliable identifications. Help is available in the form of published recordings of known species. Probably the most useful is *The Bat Detective*, a CD and accompanying booklet produced by Brian Briggs and David King. It has recordings of the British species and helpful notes and explanations that will ease the problems of recognising the different sounds produced by the bat detector. Attending bat walks or surveys organised by your local Bat Group will also be invaluable because they will take you to sites with abundant bats and with a range of species, and you will learn faster from more experienced listeners. You will also get the chance to try out a bat detector before committing a significant amount of money to buying one yourself.

The information given here should help you to get started. There are two common misconceptions that first need to be dealt with – that each species makes a unique call and that each species can be recognised because it calls on a different frequency. Both of these are true to only a very limited extent because few bat calls are territorial. Territorial calls are signals to others, which means that each species needs a distinctive call that displays its identity very clearly to other animals (if not always quite so clearly to human observers). Calls that are not signals to other bats do not need to be different.

The majority of calls made by bats are echolocation calls. Bats emit the calls and listen to the echoes in order to get information about their environment as they fly around in the dark. They are so good at this that they can catch insects in the air in total darkness. The calls are therefore finely adjusted to carry out this task in the conditions that the bat finds itself. The calls need to be different according to whether the bat is in an open environment or in dense woodland, and there is a tendency for different species to produce very similar calls in similar environments. This makes it very difficult to identify some species, particularly in cluttered environments such as woodland. The challenge of identifying bats from their calls is therefore very different from that involved in identifying, say, the territorial songs of birds or grasshoppers.

IDENTIFYING BATS WITH HETERODYNE DETECTORS

Species	*Loudest Frequency/kHz*	*Identifiable*	*How?*	*Notes*
Common Pipistrelle	45	Yes, to species	***Slaps*** loudest at 45, ***clicks*** at 55	Flittering pattern of flight in both species.
Soprano Pipistrelle	55	Yes, to species	***Slaps*** loudest at 55, almost inaudible at 45	
Nathusius's Pipistrelle	38–40	To species ?	***Slaps*** loudest at or below 40	Needs confirmation from sonogram or other evidence.
Noctule	18–25	Yes, to species	Very loud ***slaps*** at 18–20, alternating calls ('chip-chop')	Confirmed by narrow wings and high steady flight, with occasional swoops. Cannot be distinguished from each other or Serotine in cluttered environments.
Leisler's	25–30	Yes, to species	Loud ***slaps*** at 25. Almost inaudible at 18–20. No clear pattern of alternating calls.	
Serotine	30–35	Yes, to species	Characteristic syncopated rhythm of ***slaps***, single type of call	Confirmed by broad wings (like a big pipistrelle), and lower flight while feeding.

Clicks, ticks and tocks	*Myotis* bats (Whiskered, Brandt's, Daubenton's, Natterer's and Bechstein's)	30–50	To genus only (but see Daubenton's below)	Rapid regular sharp ***clicks*** or ***tocks*** – does not sound like slaps at any frequency	**If bat seen**, volume and speed of clicks can help – Natterer's rapid & quiet ("thumb on a comb") Daubenton's/Whiskered/ Brandt's louder and less rapid But **never reliable in woodland.**
	Daubenton's	30–45	Yes, too species	Very regular and moderately loud series of ***clicks***	But **only if seen and heard over water** – flies like 'a miniature hovercraft'
	Barbastelle	33	Yes, too species	***Tocks*** – heavy, moderately loud slow rattling ("castanets")	Distinctive once heard, but needs experience for confident identification.
	Brown Long-eared/ Grey Long-eared	40-45	To genus (but can be confused with Natterer's)	Almost inaudible rapid ***ticks***. Occasionally makes louder ***clicks***	Very quiet and infrequently heard. Very manoeuvrable flight, occasionally hovering.
Warbles	Greater Horseshoe	80–85	To species	Musical ***warbles*** with narrow frequency range – about 80 kHz	May be heard at half frequency.
	Lesser Horseshoe	110	Yes, to species	Musical ***warbles*** with narrow frequency range – about 110 kHz	May be heard at half frequency.

Different species of bats call at different frequencies to some extent, but the calls that bats make in cluttered environments, or when closing on prey, need to include a range of different frequencies. Most echolocation calls therefore sweep down from a high to a lower frequency, and will be audible over a range of frequency settings on a bat detector. A further complication is that a very loud call can overload the bat detector and may be heard at almost any frequency setting. The most useful piece of information in getting a clue to the species is the frequency at which the sound is loudest. Together with the sound 'quality', this feature of the call enables the practised observer to identify many bats to species level. Practice involves getting the information rapidly before the bat has flown out of range, and also learning to distinguish different 'qualities' of sound.

The first barrier to overcome when using a bat detector is distinguishing bats from other sources of ultrasound or electronic interference. Birds, insects (particularly Orthoptera and moths), electronic devices (including other bat detectors), car and burglar alarms, electric fences, running water, raindrops, clothing and zips, keys and other metallic objects, and gravel & vegetation when walked on can all produce sounds on a bat detector. Perhaps the worst distraction is any mobile phone that is switched on. It is quite impossible to use a bat detector while someone close to you is making or receiving a call, and any phone that is switched on will produce irritating *beedly-beedly-beep* noises from time to time, even when not in use. Experience will teach you which of the plethora of noises emanate from bats and which from some kind of distraction.

Having eliminated the irrelevant sounds, we are left with a range of noises that are due to bats. These can be described as *slaps*, *clicks*, *ticks*, *tocks* and *warbles*. Before describing those various qualities of sound, it is worth mentioning bat social calls, which you will certainly hear sooner or later. There is a bewildering variety of these, although many are only audible when the detector is tuned to lower frequencies than you will usually use. By far the commonest are the social calls made by pipistrelles (all three species), which sound rather like interference on a radio – a kind of loud *krrrkkkkk!*, lasting a second or so and audible with the detector tuned to 20–35kHz. Being social calls, they are species-specific, but they can only be distinguished by examining the sonograms produced from a time-expansion detector. They are nonetheless useful to alert you to the fact that pipistrelles are around when, by tuning up to higher frequencies, you will be able to confirm this and probably identify the species.

When using a bat detector, you should rapidly change the frequency up and down. You will then be able to establish (especially if listening through headphones) whether the sound is loudest at one particular frequency setting, or whether there is little change. If there is a frequency that gives the loudest sound, it will probably have a 'wet' or 'slappy' sound quality, suggesting someone being slapped around the face with a wet fish. These slaps narrow down the possibilities to six species with different loudest frequencies. These are the three pipistrelles, the Serotine, the Noctule and Leisler's Bat. Tuning the detector up to a higher frequency should cause the sound to become rather quieter clicks.

If, on the other hand, there is no dramatic change in volume and the sound on the detector is a short *click*, whatever frequency the detector is tuned to, there are eight possible species to consider. These are the *Myotis* bats (Daubenton's, Natterer's, Brandt's, Bechstein's and the

Whiskered Bat), the two long-eared bats, and the Barbastelle. The Barbastelle can be separated from the others because the sounds produced are usually rather loud and heavy clicks (*tocks*) with an erratic rhythm, and audible only when the detector is tuned between 30 and 45 kHz. Long-eared bats usually produced very quiet but rapid *ticks*, which you can hear only when the bat is very close (within 3m or so). The Myotis bats are undoubtedly the most difficult to identify: their sound is audible over a wide range of frequencies (below 30 to above 70 or 80 kHz) but the sound quality is variable. Usually the sounds are *clicks*, but sometimes they can be *ticks* or *tocks*, depending on the species and the situation. Without additional information, you will usually be unsure of the exact species.

The two species of Horseshoe Bats have calls quite unlike any other bat sounds. With the detector tuned to the correct frequency (about 80 kHz for Greater Horseshoe and about 110 kHz for Lesser), you will hear an improbable sound that has the quality of a rather strange, almost musical warble. It is quite unlike any of the other bat sounds, and makes identification of these species (once you have found them!) very easy.

SUMMARY OF SOUND QUALITIES PRODUCED BY DIFFERENT GROUPS OF BATS

Sound quality	Species	Notes
Slaps	Pipistrelles, Noctule, Leisler's, Serotine	Becomes clicks if tuned above loudest frequency
Clicks	All except the horseshoe bats	May become slaps or tocks if the frequency setting is changed
Ticks	Long-eared bats, quieter calls from Myotis bats	
Warbles	Horseshoe bats	

KEY SITES

South-west England

Arlington Court, Devon – **Lesser Horseshoe Bat**
Bracket's Coppice, Dorset – **Bechstein's Bat**
Buckfastleigh, Devon – **Greater Horseshoe Bat**
Exmoor National Park, Devon/Somerset – **Barbastelle, Bechstein's Bat**
Gordano Valley NNR, Avon – **Greater Horseshoe Bat, Serotine**
Haldon Woods, Devon – **Brown Long-eared Bat**
Isles of Scilly – **Common Pipistrelle**
Mells Valley, Somerset – **Greater Horseshoe Bat**
Minster Church, Cornwall – **Greater Horseshoe Bat**
Prideaux Wood, Cornwall – **Greater Horseshoe Bat**
Savernake Forest, Wiltshire – **Brandt's Bat, Daubenton's Bat, Natterer's Bat, Whiskered Bat, Brown Long-eared Bat**
Start Point and Berry Head to Sharkham Point NR, Devon – **Lesser Horseshoe Bat, Greater Horseshoe Bat**
Wareham Forest, Dorset – **Serotine, Natterer's Bat**

West Horrington, Somerset – **Serotine**
Willsbridge Valley, Avon – **Greater Horseshoe Bat, Noctule**

South-east England
Beddington Park, London – **Serotine, Noctule, pipistrelles**
Bedfont Lakes Country Park, London – **Nathusius's Pipistrelle**
Berwick Ponds, London – **Serotine, Noctule, pipistrelles**
Bowdown Woods, Berkshire – **Brown Long-eared Bat, pipistrelles**
Briddlesford Woods, Isle of Wight – **Barbastelle, Bechstein's Bat**
Bushy Park, London – **Brown Long-eared Bat, Daubenton's Bat**
Carisbrooke Mill Pond, Isle of Wight – **Daubenton's Bat, Lesser Horseshoe Bat**
Chiddingfold Forest, Surrey – **Bechstein's Bat**
Ebernoe Common, West Sussex – **Barbastelle, Bechstein's Bat, Grey Long-eared Bat**
Grand Union Canal, London – **Daubenton's Bat, Noctule, pipistrelles**
Highgate Wood, London – **Leisler's Bat, Natterer's Bat**
Mole Gap to Reigate Escarpment, Surrey – **Bechstein's Bat**
Mottisfont Abbey, Hampshire – **Barbastelle**
New Forest, Hampshire – **Bechstein's Bat, Brandt's Bat, Serotine**
Singleton and Cocking Tunnels, West Sussex – **Bechstein's Bat**
The Mens, West Sussex – **Barbastelle**
Wanstead Flats, London – **Leisler's Bat, Noctule, Daubenton's Bat**
Wimbledon Common, London – **Noctule, Serotine, Daubenton's Bat, Brown Long-eared Bat, pipistrelles**
Windsor Great Park, Berkshire/Surrey – **Serotine**

East Anglia
Breckland/Thetford Forest, Norfolk/Suffolk – **Barbastelle, Leisler's Bat**
Cringleford Marsh, Norfolk – **Noctule**
Eversden Wood, Cambridgeshire – **Barbastelle**
Hangman's Wood and Deneholes, Essex – **Natterer's Bat**
Wimpole Hall, Cambridgeshire – **Serotine, Barbastelle**

Midlands
Ampthill Park, Bedfordshire – **Common Pipistrelle, Soprano Pipistrelle, Natterer's Bat**
Barnwell Country Park, Northamptonshire – **Whiskered Bat**
Boarstall Duck Decoy, Oxon – **Whiskered Bat**
Bourne Woods, Lincolnshire – **Natterer's Bat, Whiskered Bat, Leisler's Bat**
Bucknell's Wood, Northamptonshire – **Noctule, Brown Long-eared Bat, pipistrelles**
Cannock Chase, Staffordshire – **Daubenton's Bat**
Cannon Hill Park, Moseley, West Midlands – **Daubenton's Bat**
Elvaston Castle Country Park, Derbyshire – **Daubenton's Bat**
Everdon Church, Northamptonshire – **Natterer's Bat**
Fawsley Lakes, Northamptonshire – **Whiskered Bat, Brown Long-eared Bat**
Fermyn Woods Country Park, Northamptonshire – **Natterer's Bat, Whiskered Bat**

Grand Union Canal, Northamptonshire – **Daubenton's Bat**
Hartsholme Country Park and Swanholme Lakes SSSI, Lincolnshire – **Common Pipistrelle, Soprano Pipistrelle**
Montford Bridge, Shropshire – **Noctule, pipistrelles**
Over Hospital, Gloucestershire – **Lesser Horseshoe Bat**
Sherwood Forest NNR, Nottinghamshire – **Noctule**
Stockgrove Country Park, Bedfordshire/Buckinghamshire – **Noctule, Daubenton's Bat**
Thornborough Bridge, Buckinghamshire – **Common Pipistrelle, Soprano Pipistrelle, Daubenton's Bat**
Tring Reservoirs, Buckinghamshire/Hertfordshire – **Noctule, Serotine, Daubenton's Bat**
Woburn Abbey and Park, Bedfordshire – **Daubenton's Bat**
Woodchester Park, Gloucestershire – **Greater Horseshoe Bat, Lesser Horseshoe Bat, Daubenton's Bat**
Wye Valley and Forest of Dean, Gloucestershire/Monmouthshire – **Greater Horseshoe Bat, Lesser Horseshoe Bat**

Northern England

Burrs Country Park, Bury, Lancashire – **Daubenton's Bat**
Carlisle Park, Morpeth, Northumberland – **Daubenton's Bat**
Corbridge, Northumberland – **Natterer's Bat, Noctule**
Jumbles Reservoir and Country Park, Bolton, Lancashire – **Daubenton's Bat, Noctule, Brown Long-Eared Bat**
Leighton Moss RSPB, Lancashire – **Noctule**
Low Catton, East Yorkshire – **Natterer's Bat**
Malton NR and Picnic site, Co. Durham – **Brandt's Bat, Noctule, Daubenton's Bat**
Middleton-in-Teesdale, Co. Durham – **Whiskered Bat**
Pennington Flash Country Park, Greater Manchester – **Noctule, Daubenton's Bat, Brown Long-Eared Bat, Natterer's Bat**
Raby Castle, Co. Durham – **Whiskered Bat**
Rothbury, Northumberland – **Daubenton's Bat, Natterer's Bat**
Shibdon Pond LNR, Co. Durham – **Noctule**
St Hilda's Church, Ellerburn, North Yorkshire – **Natterer's Bat, Whiskered Bat, Brown Long-eared Bat, pipistrelles**
Wykeham Forest, North Yorkshire – **Brown Long-eared Bat, pipistrelles**
Yarrow Valley Park, Lancashire – **Noctule, Whiskered Bat, Daubenton's Bat**

Wales

Coed y Brenin, Gwynedd – **Lesser Horseshoe Bat, Whiskered Bat**
The Lake, Llandrindod Wells, Powys – **Daubenton's Bat, Noctule, pipistrelles**
North Pembrokeshire woodlands – **Barbastelle**
Pencelly/Pengelli Forest, Pembrokeshire – **Brown Long-Eared Bat, Barbastelle, Natterer's Bat**
Stackpole Estate, Pembrokeshire – **Greater Horseshoe Bat, Whiskered Bat, Daubenton's Bat**
Wentwood Forest, Gwent – **Noctule**

Scotland

Falls of Clyde, Clyde – **Natterer's Bat**

Loch Lomond and Trossachs National Park, Clyde – **Brown Long-eared Bat**

Penninghame Pond, Newton Stewart, Dumfries & Galloway – **Daubenton's Bat, Leisler's Bat, Soprano Pipistrelle, Common Pipistrelle**

Ireland – Lesser Horseshoe Bat, Leisler's Bat, Nathusius's Pipistrelle

STATUS AND DISTRIBUTION

All 16 breeding species in Britain and Ireland are native. An additional species, the Greater Mouse-eared Bat, was declared extinct in Britain in 1991. Since then, there have been occasional records, which could be due to vagrant animals, or could mean that this species is still breeding here in very small numbers. These 17 species represent two families of bats. There are two species of Horseshoe Bats (Rhinolophidae) and 15 species of Vesper or Plain-nosed Bats (Vespertilionidae). In addition to the breeding species at least another eight species have occurred as vagrants.

Few people could have predicted that there would be a new species of common mammal recognised in the 1990s in Britain, one of the best-studied parts of the world. One of the 17 species is not only newly identified, but is also the second most abundant species of bat in the British Isles. The increased use of good quality bat detectors gave rise to the observation that Pipistrelles were sometimes heard most loudly at 45 kHz, and sometimes at 55 kHz. Detailed research work showed that the explanation was that there are two 'cryptic species', now known as the Common Pipistrelle *Pipistrellus pipistrellus*, and the Soprano Pipistrelle *P. pygmaeus*. The two are similar in general appearance but differ significantly in their DNA and ecology. They do not roost together or interbreed. From the point of view of the bat-watcher in Britain or Ireland, these two species are the ones that you are most likely to see.

The list of species may still not be finalised. The Whiskered Bat *Myotis mystacinus* and Brandt's Bat *Myotis brandtii* were only recognised as different species in the 1970s, due to the fact that they are not morphologically distinguishable. In Europe, recent research has led to the discovery of a third species from this complex, *Myotis alcathoe*. This species has not yet been recorded in Britain, but it may be found here in the future.

Because they are mammals with a high metabolic rate when active and are completely dependent on insect food, bats are very sensitive to differences in temperature. Many species are at the northern limit of their range in Britain. There are more species found in the extreme south of Britain, with fewer species the further north you go. This does not necessarily mean, however, that the numbers of individual bats will be lower further north. Brandt's Bat appears to be more common in the west and north of England than in the south, and the Soprano Pipistrelle is probably more common in the north than in the south. Certainly the largest bat roosts known in Britain are Soprano Pipistrelle roosts in Scotland, which may contain several thousand bats in a single roost.

It will be interesting to see if global warming produces changes in bat distribution in the next few years. Bats, being much more mobile than other mammal species, might be expected to expand their range northwards quite quickly as the temperature rises. There are some indications that this may already be happening. For example, there have been some recent records of Serotines a little further north than previously recorded. On the other hand, the interaction of ecological factors is always complex, and the outcome of change is difficult to predict with certainty. Since bats are near the top of the food chain, it is quite possible that changes in vegetation and insect populations will affect bats in unforeseen ways, and it is equally possible that some species might decline in numbers or contract their ranges in response to changes in food supply or to competition.

Although all the British and Irish species are also present in mainland Europe, where some of them have been shown to migrate considerable distances (1,000km or more) in spring and autumn, there seem to be only modest seasonal movements of bats in Britain. For example, whereas Barbastelles fly from a large area of central Europe to hibernate in the disused tunnels at Nietoperek in Poland, British Barbastelles hibernate in or very close to the woodlands they use in summer. Further research may well change this picture, but at present the only species which seems likely to migrate significantly in Britain and Ireland is Nathusius's Pipistrelle. This species, which has been identified only recently as resident, has produced a thin scattering of records in England and Ireland, with very few known roosts. This makes sense if it follows its European habit of migrating regularly, but it is likely to be under-recorded because it is difficult to distinguish from the Common Pipistrelle. New information is constantly emerging but, for the present, Altringham's recent book *Bats: Biology and behaviour* gives a good general guide to distribution, and Phil Richardson's *Distribution Atlas of Bats in Britain and Ireland* gives a more detailed picture up to 2000.

In the table on page 42, population estimates are given for the United Kingdom (Great Britain and Northern Ireland). These are taken from the first report of the Tracking Mammals Partnership. The estimates reflect detailed work carried out by a number of people over some years, and refined by the results obtained from the National Bat Monitoring Programme organised by the Bat Conservation Trust. A similar organisation for the Republic of Ireland (Bat Conservation Ireland) has recently been set up, but no comparable population estimates are yet available for the Republic. British Status refers to England, Scotland and Wales. Irish Status refers to the island of Ireland, i.e. the Republic of Ireland and Northern Ireland. In the table, you will notice that the horseshoe bats are listed as endangered, while several other species (Bechstein's Bat, Serotine, Grey Long-eared Bat and Barbastelle) have lower estimated numbers. The horseshoe bats have suffered a very large decline in both numbers and range in the last hundred years. The other species are all at the northern limit of their European range in Britain, but they are not thought to be endangered in European terms.

STATUS AND DISTRIBUTION

Greater Horseshoe Bat
British Status South-west England and south Wales – rare & endangered
Irish Status Absent
UK Estimate >6,600

Lesser Horseshoe Bat
British Status South-west England, Wales & the Welsh borders – rare & endangered
Irish Status South-west – locally fairly common
UK Estimate 18,000

Whiskered Bat
British Status Widespread in England, Wales & southern Scotland, but locally distributed
Irish Status Throughout – fairly common
UK Estimate 64,000

Brandt's Bat
British Status Widespread in England & Wales – more common in the north & west
Irish Status Recently discovered – status uncertain
UK Estimate 30,000

Natterer's Bat
British Status Widespread except for northern Scotland – fairly common
Irish Status Widespread
UK Estimate 148,000

Bechstein's Bat
British Status Southern – from Sussex and Surrey to Somerset and south-east Wales – very rare
Irish Status Absent
UK Estimate >1,500

Daubenton's Bat
British Status Widespread throughout England, Wales & Scotland – common
Irish Status Widespread – fairly common
UK Estimate 560,000

(Greater) Mouse-eared Bat
British Status Extreme south – extinct?
Irish Status Absent
UK Estimate ?

Serotine
British Status Widespread south-east of a line from the Wash to south-east Wales, but locally distributed
Irish Status Absent
UK Estimate 15,000

Noctule
British Status Widespread in England, Wales & southern Scotland – uncommon
Irish Status Absent
UK Estimate 50,000

Leisler's Bat
British Status Throughout England & in south-west Scotland – scarce
Irish Status Widespread & very common
UK Estimate 28,000

Common Pipistrelle
British Status Widespread throughout England, Wales & Scotland – very common
Irish Status Widespread & common
UK Estimate 2,430,000

Soprano Pipistrelle
British Status Widespread throughout England, Wales & Scotland – very common
Irish Status Widespread & common
UK Estimate 1,300,000

Nathusius's Pipistrelle
British Status Patchily distributed in England – rare
Irish Status Patchily distributed – some large colonies in Northern Ireland
UK Estimate 16,000

Barbastelle
British Status Thinly spread in southern & central England & Wales – rare
Irish Status Absent
UK Estimate 5,000

Brown Long-eared Bat
British Status Widespread throughout England, Wales & Scotland – common
Irish Status Widespread & common
UK Estimate 245,000

Grey Long-eared Bat
British Status Southern England between Somerset & West Sussex – very rare
Irish Status Absent
UK Estimate 1,000

VAGRANT BATS

Particoloured Bat *Vespertilio murinus*
This is a highly migratory species that breeds in Eastern Europe and southern Scandinavia. Northern populations fly to southern Europe to hibernate. There are recorded movements of as far as 900km. This is the most regularly recorded vagrant bat in Britain and Ireland, with at least seven records in the past 100 years.

Northern Bat *Eptesicus nilssoni*
The Northern Bat has a widespread distribution from eastern France through central Europe and Scandinavia east to Siberia, Tibet, Japan and northern India. Although not really migratory at least two have been recorded in the British Isles: an individual picked up in Surrey in 1987 and another on an oil-rig off Aberdeen in 1993 (Macdonald & Tattersall, 2001).

Kuhl's Pipistrelle *Pipistrellus kuhlii*
Kuhl's Pipistrelle is an uncommon resident on the island of Jersey. Away from here, this species is common in North Africa, southern Europe and east to Pakistan. It is thought to be extending its range northwards. There are at least seven British records, the last of which was on the Isle of Wight in September 2002. At least four of the records stem from accidental importations.

Savi's Pipistrelle *Pipistrellus savii*
Savi's Pipistrelle is distributed throughout southern Europe, north-west Africa and Asia. It is probably migratory, with three British records up to 2001.

Hoary Bat *Lasiurus cinereus*
The Hoary Bat is native to North America and is an exceptionally rare vagrant to Europe with two records from the Scottish islands.

Silver-haired Bat *Lasionycteris noctivagans*
This North American species has been recorded twice in Britain between 1980 and 1999. On both occasions they arrived by artificial means. One was found in the cockpit of a US plane and the other was located amongst some imported wood.

Mexican Free-tailed Bat *Tadarida brasiliensis*
In 2003, an individual of this species was located in Kent. Cooperation between Britain and the US eventually led to its safe return to North America.

European Free-tailed Bat *Tadarida teniotis*
A very rare vagrant from Mediterranean Europe. In March 2003 an individual was found in a Cornwall churchyard and was immediately taken into care. It was the first record of this species in this country for 17 years.

LAGOMORPHS

(European) Rabbit *Oryctolagus cuniculus*
(European) Brown Hare *Lepus europaeus*
Mountain Hare *Lepus timidus*

The three species of lagomorph found in Britain and Ireland are among our most conspicuous mammal species. All three are diurnal to some degree, are sufficiently large to be easy to find and are found in sizeable numbers in prime habitat, although there has been recent evidence of a significant decline in the Brown Hare population.

IDENTIFICATION

All three species, although superficially similar, should be easy to identify in the field.

Rabbit Separated from hares by its smaller size, shorter ears (lacking black tips) and shorter hind legs in relation to the body. Tail all white.

Brown Hare Easily told from Rabbit by its larger size and longer limbs, its longer black-tipped ears and the black upper tail. Much yellower overall than Mountain Hare which tends to be greyish-brown in summer.

Mountain Hare Smaller and greyer than Brown Hare with shorter black-tipped ears and an all-white tail. The long limbs typical of hares help distinguish the species from Rabbit. The coat begins to change colour in September, when the brown of summer turns to bluish-grey. It then continues to fade and the white winter coloration is usually attained by December. In spring, individuals often retain a good deal of their winter colouring until April. Irish Mountain Hares tend to be brighter than British animals and can be confused with introduced Brown Hares. Irish animals do not turn white in winter.

SEEING LAGOMORPHS

Rabbits are easy to see and can be observed throughout the day particularly in areas with little disturbance. Rabbit warrens provide opportunities to observe the species at close range and can also be good places to see their predators, e.g. Red Foxes and Stoats.

Brown Hares spend much of the day in cover and are easiest to see at dusk when they emerge to feed. They are easy to see from September to March when there is less cover, and are particularly visible during the mating season, March–April, when they engage in chases and boxing activity.

Mountain Hares are usually easy to see throughout the winter when their white winter coat is conspicuous against the frequently snowless backdrop. They are also conspicuous in April when they are very active and, at the same time, comparatively tame. In some areas such as the southern Pennines they are relatively easy to find throughout the year. Although its name suggests that a tortuous hike may be required to see this species, in reality this is not the case at all. In fact, in many areas, it is not even necessary to leave the warmth of your car. Once suitable habitat is located, it is advisable to scan along the tops of hills and ridges where the shapes of Mountain Hares can often be seen against the skyline. Look particularly for areas of short turf as the hares are usually more conspicuous there than they are when camouflaged against a backdrop of old, tall heather.

KEY SITES

A list of sites is not really necessary for Rabbits or Brown Hares but the following locations are reliable sites for observing **Mountain Hares**.

Northern England

Pennines/Peak District, South Yorkshire/Derbyshire

Scotland

Findhorn Valley, Highland
Glen Affric, Highland
Lynemore, Highland
Outer Hebrides
Shetland
Slochd Summit, Highland
The Cairngorms, Highland

STATUS AND DISTRIBUTION

Rabbit
Population 40,000,000
Range Occurs throughout Britain and Ireland at all altitudes up to the tree line. Occurs on most offshore islands, where inbreeding leads to high levels of melanism.
Habitat Rough ground and pasture, hedgerows and where inbreeding leads to high levels of melanism. woodland edges. Exhibit preference for areas with light sandy soils.

Brown Hare
Population 817,500
Range Occurs throughout the British Isles including Orkney and the Outer Hebrides but absent from the north-west and western Highlands. Introduced to Ireland where it still occurs in the north-west.
Habitat Primarily found in arable farmland and pasture interspersed with hedgerows and woodland where they shelter during the day.

Mountain Hare
Population 360,000 in Britain. 82,000 in Northern Ireland
Range Common in the Scottish Highlands and locally in southern Scotland and the Hebrides and Pennines. Populations in southern Scotland and the Pennines are introduced. Widespread in Ireland but declining in the north.
Habitat Heather moors and pasture in Britain (habitat shared with Rabbits and Brown Hares). Widespread in Ireland where particularly abundant on farmland.

SQUIRRELS

(Eurasian) Red Squirrel *Sciurus vulgaris*
(Eastern) Grey Squirrel *Sciurus carolinensis*

Unfortunately our only native squirrel, the Red, is gradually being ousted from its woodland haunts by Grey Squirrels. Originally introduced in Cheshire in 1876 as an ornamental species the Grey Squirrel has colonised much of the British Isles, and the current distribution of the Red Squirrel closely reflects the areas where Greys do not occur. However, although there is evidence that the decline of Red Squirrels has been greatest in areas where Greys occur there is also evidence that habitat destruction and disease, particularly coccidiosis and parapoxvirus, have also contributed to the decline. The remaining strongholds of Red Squirrels probably still exist because of the lack of woodland corridors by which Greys can reach them. Unfortunately, in 2003 there were 342 verified sightings of Greys in the Merseyside area (a significant increase from the 76 verified sightings the previous year) and by November 2003 at least nine Red Squirrels there had contracted parapoxvirus (Macdonald & Tattersall, 2004). Fortunately proposals to cull Grey Squirrels in areas where they now pose a threat to remaining populations of Reds were announced in early 2006.

IDENTIFICATION

Red Squirrel	Smaller and more compact than Grey Squirrel, with tufty ears. Coloration can vary from deep brown through red and chestnut to grey-brown. Continental animals may also be blue-grey, grey or black.
Grey Squirrel	May exhibit some red or chestnut coloration on the back and the legs, but generally grey although melanistic individuals are common in some areas, e.g. around Hitchin and Letchworth in Hertfordshire. Young Grey Squirrels could be confused with Edible Dormice in areas where they both exist but the nocturnal nature of the latter reduces the likelihood of confusion.

SEEING SQUIRRELS

Red Squirrels are arboreal and diurnal with peaks of activity around dawn and dusk in summer, and late morning in winter. They do not hibernate, although they may lie up in their dreys for several days during severe winter weather. They are easy to observe. Although they are predominantly arboreal, they do descend to the ground to collect food items that have fallen to the floor. Careful searching at the sites listed below should produce sightings.

Grey Squirrels are common and widespread and simply entering woodlands where they occur is likely to produce sightings. They can be very vocal in the autumn making locating them even easier.

KEY SITES

Grey Squirrels are sufficiently common that specific sites are not needed. The sites listed below still hold populations of Red Squirrels.

South-west England
Brownsea Island, Dorset

South-east England
Briddlesford Woods, Isle of Wight
Isle of Wight woodlands

East Anglia
Breckland/Thetford Forest, Norfolk/Suffolk

Northern England
Formby Point NT, Merseyside
Grizedale Forest Park, Cumbria
Kielder Forest, Northumberland
Rothley Lakes, Northumberland

Wales
Coed y Brenin, Gwynedd

Scotland
Abernethy Forest, Highland
Ardcastle, Argyll & Bute
Clashindarroch, Aberdeenshire
Culbin, Moray
Glen Urquhart, Highland
Kilmun Arboretum, Argyll & Bute
Mabie Forest, Dumfries and Galloway
Plodda, Highland
The Durris Forest, Aberdeenshire

STATUS AND DISTRIBUTION

Red Squirrel
Population 161,000* of which 121,000 are in Scotland
Range Scottish Highlands, north-east England, north Cumbria, and a few parts of North Wales. Small populations on the Isle of Wight, on three islands in Poole Harbour and around coastal Lancashire and Merseyside.
Habitat Stable populations in Britain tend to be found in large tracts (>100ha) of relatively mature coniferous forest. They may, however, also be found in smaller woods and copses where Grey Squirrels are absent.

Grey Squirrel
Population 2,520,000*
Range Widespread throughout England, Wales, Northern Ireland and central Scotland.
Habitat Broad-leaved, mixed and mature coniferous forests. Parks, gardens and urban areas with mature trees.

* Excluding Ireland

RODENTS

RODENTS

Bank Vole *Clethrionomys glareolus*
Field Vole *Microtus agrestis*
Common (Orkney) Vole *Microtus arvalis*
(European) Water Vole *Arvicola terrestris*
(Long-tailed Field) Wood Mouse *Apodemus sylvaticus*
Yellow-necked (Field) Mouse *Apodemus flavicollis*
(Eurasian) Harvest Mouse *Micromys minutus*
(Western) House Mouse *Mus domesticus*
Brown Rat *Rattus norvegicus*
(House) Black Rat *Rattus rattus*
Hazel Dormouse *Muscardinus avellanarius*
Edible Dormouse *Glis glis*

Including squirrels, covered separately, 14 species of rodent currently breed in the UK with at least another six species having occurred at one time or another. The Coypu, once a common sight in parts of East Anglia with a population of over 200,000 in 1962, is now extinct following a campaign organised by the Ministry of Agriculture, Fisheries and Food which saw more than 120,000 animals killed between 1970 and 1987. The last animal was trapped in December 1989. In addition American Beavers, Golden Hamsters, Indian Crested Porcupines and Muskrats have all bred in a natural state in Britain but all stem from either deliberate or accidental releases and none are believed to survive in a wild state at the present time. Plans to reintroduce European Beavers into Scotland have recently been rejected by the Scottish Executive, but Beavers have been introduced to trial sites in Kent and Gloucestershire.

IDENTIFICATION

Bank Vole Distinguished from Field Vole by the reddish-brown coat of the adults that often contrasts with greyish flanks, its longer tail and more prominent ears. Both species exhibit a blunt nose and small eyes. Juveniles are more easily confused, but look for the longer tail and more prominent ears.

Field Vole A small yellowish-brown to grey-brown vole with a short tail, blunt snout and small ears and eyes. Juveniles are usually darker than adults and may be confused with the Bank Vole except that the tail is shorter in the Field Vole. Never shows the chestnut-brown coloration of the Bank Vole.

Common (Orkney) Vole Extremely close in appearance to the Field Vole but may be separated at close range by the shorter coat and less hairy ears. Fortunately, the Common Vole is the only vole species on Orkney.

Water Vole The largest vole found in Britain, which may be mistaken for a Brown Rat (which often shares its aquatic habitat). Usually dark brown (or black, especially in northern Scotland), with a typical blunt, vole-like muzzle. The tail is slightly furry and much shorter than the body. Has small ears. The Brown Rat has a pointed muzzle, more conspicuous ears and a much longer tail.

Wood Mouse Brown above and very white below with a very long tail. Most similar to the slightly larger Yellow-necked Mouse, from which it can usually be distinguished by the lack of a complete yellow collar, although care should be taken, as many do show at least some yellow on the throat. Juveniles may be mistaken for the House Mouse but note the large eyes, ears and feet.

Yellow-necked Mouse Bigger than the Wood Mouse and can usually be distinguished by the complete yellow collar that is discernible even in juveniles. The Wood Mouse can, however, show variable amounts of yellow on the throat so care is needed (although the collar is never usually complete). The contrast between the orange-brown upperparts and pale underparts is usually greater than in the Wood Mouse, and the tail is longer, and thicker at the base than in that species.

Harvest Mouse This is the smallest British rodent and easily distinguished from other mice species by its golden-brown upperparts, blunt muzzle and small, hairy ears. The bicoloured tail is sparsely haired and similar in length to the head and body.

House Mouse Dull greyish-brown and can be distinguished from Yellow-necked and Wood Mice by smaller eyes and shorter hind feet. They are also rarely as pale below as the other two species. The tail is slightly thicker and scalier than in other mice species.

Brown Rat Chunkier than the Black Rat and has proportionately smaller eyes and ears. The ears are finely furred (compared with the hairless ears of the Black Rat) and the tail is relatively shorter and thicker. When swimming the Brown Rat can easily be confused with the Water Vole, but look for the rat's pointed muzzle, prominent ears and longer tail.

Black Rat Similar in appearance to the Brown Rat but may be distinguished by the larger eyes and ears, and longer, thinner and hairless tail. The fur colour is not diagnostic as it is highly variable with certain individuals being brown, and some Brown Rats may be black. To some eyes the Black Rat more closely resembles a large House Mouse than a Brown Rat in structure.

Hazel Dormouse Adults are usually orange-brown on the back and pale buff-white on the throat. A good feature is the bushy tail, which is unusual in a mouse-sized animal. Other features to look for include the short muzzle, long black whiskers, and prominent black eyes.

Edible Dormouse Only likely to be confused with a small Grey Squirrel, which is also grey with a bushy tail. However, even the largest dormouse is considerably smaller than the smallest Grey Squirrel, and this coupled with the fact that the two species are active at completely different times of day makes confusion unlikely.

SEEING RODENTS

Although views generally tend to be brief most rodents are easiest to see in the field during the harvest in August and September when there is an abundance of food and less cover in which they can hide.

Most small rodents rarely give more than brief glimpses in the field and to see them well it is essential to catch them. The best-known small mammal trap is the expensive Longworth trap although several new cheaper traps have recently come onto the market. A range of small mammal traps can be obtained from Alana Ecology. A minimum of a dozen traps is required to achieve anything worthwhile, and consequently it may be better to find a local mammal group who go out trapping small mammals. Organisations such as The People's Trust for Endangered Species, Mammals Trust UK and The Mammal Society also run regular small mammal days or courses providing the opportunity to see a range of small mammal species in the hand.

Many species including the **Wood Mouse, Yellow-necked Mouse** and **Bank Vole** will use nest boxes in winter and checking boxes can prove fruitful, but please remember that nest boxes in areas known to hold **Hazel Dormice** can only be checked under licence. Also remember that mice can give you a nasty bite so gloves are recommended when checking boxes in winter. Bird tables frequently attract Wood Mice and Bank Voles and baiting an area with raisins, seeds and cereals will often attract rodents. See the Fieldcraft section for further details.

The best time to observe **Water Voles** is during the summer, as they spend little time out of their burrows during the colder months. They breed in July, and this is possibly the best month to look for this species as they spend up to a third of their time out of the nest. Sitting patiently along river banks where there is evidence of Water Vole activity: e.g. well-worn runs, droppings etc., will often bring its rewards and if there are Water Voles in the vicinity they can often be attracted with apple or other fruit baits.

Harvest Mice appear to be most active shortly after dusk and just prior to dawn. Trapping studies indicate that this species is more nocturnal in summer and diurnal during the winter. Trapping is by far and away the easiest way to see this species and the People's Trust for Endangered Species, in conjunction with the Mammals Trust UK, run small mammal trapping sessions where this species is likely to be seen. Observing these animals in their natural environment is extremely challenging. Their hearing is acute and they will react sharply and either freeze or dive for cover at rustling sounds as far as 7m away. Even sitting quietly and patiently in suitable habitat is only likely to produce a sighting very occasionally. Locating a nest would enhance the chances. Made from grass leaves, they are attached to various plants and grasses and can be well camouflaged in the summer and autumn but are much more obvious in winter as round balls of grass suspended amongst the dying stems. In some areas Harvest Mice will come to food set out for birds. At others it has been found that placing a tennis ball with a hole cut in its side about half a metre off the ground in suitable habitat may attract the species, especially if the ball is baited with seed.

House Mice may be encountered during daylight hours and the easiest places to observe this species are outbuildings particularly around farms. The underground stations of central London, particularly on the Piccadilly and Northern Lines, have traditionally been excellent places to see House Mice at any time of day and night, generally scurrying about under the tracks.

Following the eradication of the previously easy to see population on Lundy, **Black Rats** now pose a major challenge for anyone wishing to see this species in Britain. This is not a species you are simply going to stumble across. It will require a good deal of effort and no little luck to encounter one. Overnight stays on the Shiant Islands offer the only real chance of seeing one.

Being largely nocturnal **Hazel Dormice** are rarely encountered during the day. However, if they are they are often sluggish, providing opportunities for prolonged views. They are strongly associated with ancient woodland, especially hazel coppice, and their presence may be confirmed by the discovery of hazelnuts which show a neat, round hole with a steep, smoothly cut edge surrounded by oblique teeth marks on the nut. (Wood Mice make a less evenly rounded hole and Bank Voles make a series of cuts around the hole and tend to leave no scratchy marks.) Another useful sign that dormice are present in an area is the presence of nibbled honeysuckle flowers scattered on the ground. Hazel Dormice are a protected species under the Wildlife and Countryside Act of 1981. They cannot be handled or trapped, and their nests (nest boxes) cannot be inspected without a licence. The best way to see the species is unquestionably to join a licensed Hazel Dormouse worker on a nest search. Many local mammal groups and organisations such as The Mammal Society, Mammals Trust UK and the People's Trust for Endangered Species have regular dormouse days. The Mammal Society also runs regular courses for people wishing to obtain a licence and attending one of these courses should provide an opportunity to see the Hazel Dormouse in the wild.

In the right habitat signs of **Edible Dormouse** are easy to find. They cause tree damage by chewing bark, buds and growing shoots, and consequently the tops of many of the trees in the Edible Dormouse's range are dead or dying, often with telltale strips of dead bark hanging down. During the summer they make a nest in a tree hole and in some areas they utilise nest boxes. Actually seeing this species may, however, require a lot of hard work. Due to their nocturnal, arboreal habits it will be necessary to spend time spotlighting in suitable woodland. Edible Dormice are extremely vocal and you will soon know if there are any in the area. Having located an individual by call, try to pinpoint its location before turning on the spotlight as it will quickly disappear once the spotlight is turned on. You will increase your chances of a prolonged view if you use a spotlight with a red filter. Warm, still evenings are best, particularly between late June and mid-August. Alternatively the People's Trust for Endangered Species, in conjunction with Mammal Trust UK, runs trips most years to view Edible Dormice in the Tring area. Participants check nest boxes as part of a long-term study of the species and this probably offers one of the best opportunities for gaining good views.

KEY SITES

Many rodents are widely distributed but the following sites provide opportunities for seeing the less common **Water Vole, Yellow-necked Mouse, Harvest Mouse, Hazel Dormouse** and **Edible Dormouse**.

South-west England

Arne RSPB Reserve, Dorset – **Harvest Mouse**
Brownsea Island, Dorset – **Water Vole**
Cardinham Woods, Cornwall – **Hazel Dormouse**
Haldon Woods, Devon – **Hazel Dormouse**
Lydford, Devon – **Hazel Dormouse**
Savernake Forest, Wiltshire – **Yellow-necked Mouse**
The Fleet, Dorset – **Water Vole, Harvest Mouse**

South-east England

Abbots Wood, East Sussex – **Hazel Dormouse**
Bedfont Lakes Country Park, London – **Water Vole**
Bowdown Woods, Berkshire – **Hazel Dormouse**
Briddlesford Woods, Isle of Wight – **Hazel Dormouse**
Elmley Marshes RSPB Reserve, Kent – **Water Vole, Harvest Mouse**
Rye Harbour NR, East Sussex – **Water Vole**
Stodmarsh NNR, Kent – **Water Vole**
The Wetland Centre, Barnes, London – **Water Vole**

East Anglia

Bradfield Woods NNR, Suffolk – **Yellow-necked Mouse, Hazel Dormouse**
Cley NWT Reserve, Norfolk – **Water Vole**
Cringleford Marsh, Norfolk – **Water Vole**
Holkham, Norfolk – **Water Vole**
Strumpshaw Fen RSPB Reserve, Norfolk – **Water Vole**
Tollesbury Wick, Essex – **Water Vole**
Wicken Fen NT, Cambridgeshire – **Water Vole, Harvest Mouse**

Midlands

Ashridge Park, Hertfordshire/Buckinghamshire – **Edible Dormouse**
Cotswold Water Park, Gloucestershire/Wiltshire – **Water Vole**
Dymock Woods, Gloucestershire – **Hazel Dormouse**
Rye Meads RSPB Reserve, Hertfordshire – **Water Vole, Harvest Mouse**
Saltfleetby-Theddlethorpe Dunes NNR, Lincolnshire – **Water Vole**
Slimbridge WWT, Gloucestershire – **Water Vole**
Wendover Woods, Buckinghamshire – **Edible Dormouse**
Wyre Forest, Worcestershire/Shropshire – **Yellow-necked Mouse, Hazel Dormouse**

Northern England

Annitsford Pond LNR, Northumberland – **Water Vole**
Blacktoft Sands RSPB Reserve, Yorkshire – **Water Vole**
Leighton Moss RSPB Reserve, Lancashire – **Harvest Mouse**

Wales

Brechfa Forest, Carmarthenshire – **Water Vole, Hazel Dormouse**
Crychan, Powys – **Hazel Dormouse**

Dyfi, Gwynedd – **Hazel Dormouse**
Teifi Marshes NR, Ceredigion – **Water Vole**
Tregaron Bog/Cors Caron NNR, Ceredigion – **Water Vole**

Scotland
Outer Hebrides – **Black Rat**

STATUS AND DISTRIBUTION

Bank Vole
Population 23,000,000*
Range Widespread on mainland Britain and in southern Ireland. Also occurs on many of the offshore islands including Skomer which even has its own subspecies.
Habitat Woodland, scrub, banks and hedgerows.

Field Vole
Population 75,000,000
Range Widespread but patchily distributed throughout the British mainland. Present on many of the Outer Hebrides but absent from Orkney, Shetland, Lundy, the Isles of Scilly, Isle of Man & Ireland.
Habitat Rough ungrazed grasslands including meadows, field margins, plantations with a grassy understorey, hedgerows, dunes, blanket bogs and moorland.

Common (Orkney) Vole
Population 4,000,000
Range Mainland Orkney, Rousay, Sanday, Westray and South Ronaldsay.
Habitat Old peat cuttings, rough grasslands, drainage ditches and roadside verges

Water Vole
Population 875,000
Range Found throughout Britain, but absent from Ireland and most of the islands except Anglesey and the Isle of Wight. Its distribution is now patchy, and populations in many areas have declined at an alarming rate. It is being reintroduced into some areas.
Habitat Grassy banks along slow-flowing rivers, lakes, ponds and marshland. Reedbeds appear to support large populations.

Wood Mouse
Population 38,000,000*
Range Widespread on the British and Irish mainland. Absent from only very small islands e.g. Lundy, Isle of Man, North Rona & Isles of Scilly except Tresco & St Mary's.
Habitat Ubiquitous, occurring in most habitats including houses.

Yellow-necked Mouse
Population 750,000
Range Principally south-east England, East Anglia and along the Welsh borders with isolated pockets in south-west England.
Habitat Ancient deciduous woodland, orchards, wooded gardens and hedgerows. Often found in rural houses in winter.

Harvest Mouse
Population 1,425,000
Range Commonest in southern England and along the coastal belt of Wales with scattered colonies possibly resulting from introductions elsewhere.
Habitat Dense vegetation such as long grasses at the base of a hedge as well as brambles, cornfields and reedbeds, hedgerows, fences and drainage ditches.

House Mouse
Population 5,400,000*
Range Widespread in Britain and Ireland including most small islands.
Habitat Largely associated with urban areas but also found in hedgerows, fields and farm buildings.

Brown Rat
Population 6,800,000*
Range Widespread in Britain and Ireland and on offshore islands including unpopulated islands.
Habitat Found in farms, rubbish tips, sewers, urban waterways and warehouses. Also occurs in hedgerows and arable fields containing cereal crops.

Black Rat
Population 230-400
Range Now restricted to the Shiant Islands in the Outer Hebrides having been eradicated from Lundy. Odd individuals turn up sporadically on the mainland, normally in the vicinity of ports.
Habitat Although generally associated with human habitation in most of the world, British animals tend to be found on rocky shores and cliffs.

Hazel Dormouse
Population 40,000
Range Commonest along the south coast from Cornwall to Kent and roughly as far north as a line drawn between the Thames and the Severn although it is also found in parts of Wales and sparsely in parts of the Midlands. In northern Britain it is restricted to three sites in Cumbria and Northumberland. Reintroduction schemes are under way in at least nine counties.
Habitat Mature, mixed deciduous woodland, scrub and hedgerows.

Edible Dormouse
Population 10,000
Range Now common in the Chilterns in Buckinghamshire, Hertfordshire and Berkshire, having been introduced at Tring in 1902. It is roughly confined to a 500km^2 triangle bounded by Beaconsfield, Ivinghoe and Aylesbury. It occurs in particularly dense populations in and around the towns of Wendover and Amersham and adjoining Chesham Bois.
Habitat Extensive and well-interconnected mixed woodland including beech and conifer trees, particularly larch and spruce.

* Excluding Ireland

CARNIVORES

Red Fox *Vulpes vulpes*
(Eurasian) Badger *Meles meles*
(Eurasian) Otter *Lutra lutra*
(European) Polecat *Mustela putorius*
Feral Ferret *Mustela furo*
American Mink *Mustela vison*
Stoat *Mustela erminea*
(Least) Weasel *Mustela nivalis*
Pine Marten *Martes martes*
Wild Cat *Felis silvestris*

Containing some of our most charismatic mammals, this exciting group includes three of our most sought after mammals: the Wild Cat, Pine Marten and Polecat, whilst sightings of many of the others such as Badger and Stoat are always fulfilling. Although some are still threatened it is heartening to see that the populations of many species, including Pine Martens, Polecats and Otters, are stable or increasing. However, the Wild Cat continues to decline and the subspecies found in Britain is now one of the most endangered felids in the world.

Unfortunately, at the time of writing Defra has proposed a mass cull of Badgers over large parts of western England in an attempt to reduce the spread of bovine TB. Although the bulk of the culling would be carried out in Devon, Cornwall, Somerset, Avon and Gloucestershire, it would also take place in the West Midlands, Dorset and Sussex, and it is feared that Badgers will become an increasingly rare sight if the cull goes ahead.

IDENTIFICATION

Red Fox Discounting feral dogs, the Red Fox is the only wild dog species to be found in Britain and it is an extremely familiar mammal. Identification should be straightforward.

Badger The Badger is a very familiar robust, powerful animal and its distinctive black and white head patterning should make identification a relatively straightforward process.

Otter The Otter is relatively easy to identify. Its predominately aquatic habits mean that the only confusion risks are American Mink and seals. Its larger size should readily distinguish it from the American Mink even at range. Inexperienced observers may confuse this species with seals, but look for the uniform brown coat and long tapering tail.

Polecat / Feral Ferret Most likely to be confused with American Mink, the Polecat should be easily distinguished from this species if good views are obtained. Look for the creamy underfur that covers most of the body, the prominent facial mask and the white ear margins. Feral Ferrets pose a com pletely different problem and it is likely that some Polecat-

Ferret hybrids are indistinguishable from true Polecats on external features.

American Mink Most wild mink are dark-brown (appearing black at a distance) with a white chin and often white patches on the chest, belly and groin. The tail is slightly bushy and approximately half the length of the body. Although it can be confused with the Otter by the inexperienced it is much smaller than that species which also has a smoother coat and a longer, thicker tail. It is more likely to be confused with the Polecat or Feral Ferret.

Stoat One of only two British mammals to turn white during the winter months. This alternative pellage is called ermine, and is usually found in the northern parts of Britain and Ireland. Stoats at lower latitudes may also turn white in response to particularly cold conditions. The black tip to the tail is always retained and along with the normally noticeably larger size is the best clue to separating it from the closely related Weasel.

Weasel The Weasel is easily distinguished from all other mammals other than the Stoat. It is smaller than a Stoat with a relatively short tail lacking a black tip. Weasels do not turn white in winter in Britain.

Pine Marten Easily distinguishable from other Mustelids. It possesses a characteristic loping gait, a cream or orange throat and chest patch, a long, bushy tail, large ears and more upright body carriage than feral cats or Mustelids. (At least two American Martens have been found dead in Northumberland. These individuals, which were likely to be fur-farm escapees, were identified through DNA analysis and observers should note that field separation of these two closely related species is impossible.)

Wild Cat If seeing a Wild Cat was not hard enough the identification of this species is extremely difficult due to the presence of feral cats and hybrids within its range. A good view is essential and even then it is debatable whether some individuals can be positively identified. A recent DNA-based study has identified five characteristics diagnostic of a pure 'Scottish' Wild Cat:

1. The four nape stripes are distinct, broad and wavy.
2. The dorsal stripe always ends at the base of the tail whereas on feral cats it often continues to the tip of the tail.
3. The rump and flanks are unspotted.
4. The tail is bushy with a blunt, broad black tip and two or three additional black bands.
5. There are no white patches on the feet.

SEEING CARNIVORES

Red Foxes may be encountered almost anywhere and are most often seen at night, although dawn and dusk often offer a better chance of prolonged views. Watching active dens in spring can be particularly rewarding when the cubs emerge to play towards dusk. The habit of feeding foxes in gardens has become more popular as they have become increasingly common in urban areas. However seeing foxes in a more natural environment is far more rewarding.

Badgers are common in Britain and Ireland but their nocturnal habits and unobtrusive nature make them easy to overlook. If a sett is located, they can be watched when they emerge to feed, particularly during the short summer nights when they are more likely to emerge in daylight. When visiting setts it is important to remain downwind as badgers will quickly disappear if they detect human scent. There are numerous sites where badgers can be viewed, often from the comfort of hides. The website of the Badger Trust (www.badger.org) provides details of where to watch badgers (see also pp. 255–259). In addition, most of the sites on pages 61–63 provide opportunities for seeing Badgers.

Given their diurnal activity, coastal **Otters** are generally easier to observe than inland populations although patience is still required. Coastal animals are particularly active on the falling tide as prey becomes exposed in rapidly shrinking pools. Many good areas for Otters can be viewed from roads along the west coast of Scotland and the outer isles and carefully scanning bays and rocky shorelines on the falling tide will often produce results. Once located you can often obtain prolonged views. Finding spraints (faeces) is a good way to determine whether Otters are in an area. Inland Otters are generally nocturnal, particularly in areas prone to disturbance. However, with luck they can occasionally be seen at some of the sites listed on pages 61–63 in the hour or two prior to dusk.

Radio tracking of **Polecats** has shown them to be predominately nocturnal with peaks of activity around dusk and dawn. It is consequently necessary to go searching for them between dusk and dawn. Recommended tactics include driving slowly around minor roads or walking paths in suitable habitat at night. This is an extremely difficult species to catch up with, and a lot of luck is required to come across one even in areas where they are common. They are most likely to be encountered during daylight in June when females are forced to hunt to feed their fast-growing litters. In winter they are known to move into farmyards to feed on rodents. No fewer than 50 were trapped on one estate in Hertfordshire in 2001, and the frequency with which road kills are now found suggests that Polecats are common in the area, yet they remain frustratingly elusive.

American Mink are not easy animals to observe. The best time to see them is during late summer, particularly August and September when the young become independent and are seemingly unafraid of people. To search out this species actively requires patience in good habitat, although chance sightings frequently occur.

Most views of **Stoats** and **Weasels** tend to be short-lived as they dash across the road in front of the car. Obtaining prolonged views is far more difficult and is often a matter of chance. If specifically looking for Stoats and Weasels it is probably best to concentrate on linear features such as hedgerows, banks and old walls, especially in autumn and winter when cover is sparser. Fortunately they are very inquisitive and can often be pished or squeaked in to obtain better views. Sitting quietly and watching Rabbit warrens for periods of time can also be rewarding. Stoats can sometimes be located by pinpointing anguished squeals of Rabbits.

Although **Pine Martens** are usually quite difficult to find, they can be seen tracking across minor roads at night or bounding across the edges of woodland clearings. They have also started to take advantage of the ready-made food source offered at bird tables and are also frequently seen at well-visited car parks and picnic sites where the animals look for scraps of food as dusk falls. Consequently baiting can prove fruitful: meat and strawberry jam both work well. They are particularly common on the Ardnamurchan Peninsula in western Scotland.

Wild Cats are without doubt the most highly sought after British mammals, yet they are rarely seen in the wild. Seeing a Wild Cat requires a lot of patience, a considerable amount of effort and a huge amount of luck. In some areas driving around spotlighting at night can produce results. In others, sitting patiently in likely looking places, such as forest edges, may prove beneficial especially in areas where human activity is minimal, and where water and prey, such as Rabbits, are plentiful. Wild Cats are known to sit over Rabbit holes and warrens waiting for their prey to emerge. They become more diurnal during late autumn as they gather food in preparation for winter. Wild Cats leave various signs around their territories and looking for signs is a good way to find out if there are any animals in the area, although it may be very difficult to distinguish between signs left by Wild Cats and those left by feral cats. Wild Cat signs include:

1. Droppings (scats) that are often left in obvious places such as in the middle of a track or on large rocks.
2. Prey remains: although such carrion is usually cleared up fairly quickly by the local scavengers.
3. Footprints in soft mud – most tracks will only exhibit four paws and the central pad.
4. Scratching posts (usually trees) that may be found at the edge of an animal's territory, although domestic cats often have these.

KEY SITES

Red Foxes are widespread and easy to see so specific sites are not necessarily appropriate. Stoats and Weasels are common throughout Britain and Ireland, and they can be found in many different habitats and locations. Specifically looking for Stoats and Weasels is difficult and most sightings are simply good luck. Many of the sites listed below are managed by conservation organisations and provide facilities for visitors.

South-west England

Offwell Woods, Devon – **American Mink**
Radipole Lake RSPB Reserve, Dorset – **American Mink**
River Camel, Cornwall – **Otter**

South-east England

Elmley Marshes RSPB Reserve, Kent – **American Mink, Stoat, Weasel**
Rye Harbour NR, East Sussex – **American Mink**
Stodmarsh NNR, Kent – **American Mink**

East Anglia

Bedford Purlieus NNR (Rockingham Forest), Cambridgeshire – **Otter**
Barton Broad NNR, Norfolk – **Otter**
Fowlmere RSPB Reserve, Cambridgeshire – **Otter, American Mink**
Minsmere RSPB Reserve, Suffolk – **Otter, Stoat, Weasel**
Strumpshaw Fen RSPB Reserve, Norfolk – **Otter**

Midlands

Cotswold Water Park, Gloucestershire/Wiltshire – **Otter, American Mink, Stoat, Weasel**
Wardley Wood, Leicestershire – **Badger**
Wendover Woods, Buckinghamshire – **Badger, Weasel**
Woodchester Park, Gloucestershire – **Badger**
Wye Valley and Forest of Dean, Monmouthshire/Gloucestershire – **Badger, Polecat, American Mink**

Northern England

Chopwell Woodland Park, Tyne and Wear – **Badger, Otter**
Kielder Forest, Northumberland – **Otter**
Mount Grace Priory, North Yorkshire – **Stoat**
River Derwent and Bassenthwaite Lake, Cumbria – **Otter**
Rothley Lake, Northumberland – **Otter**
Upper Derwent, North Yorkshire – **Otter**

Wales

Brechfa Forest, Carmarthenshire – **Badger, Polecat**
Brecon Beacons NP – **Otter**
Coastal Gwynedd – **Polecat**
Crychan, Powys – **Polecat**
Moel Famau, Denbighshire – **Polecat**
Nercwys, Denbighshire – **Polecat**
Radnor Wood, Powys – **Badger**
River Twyi/Towy, Carmarthenshire – **Otter**
Stackpole Estate, Pembrokeshire – **Otter**
Tan y Coed, Gwynedd – **Badger**
Teifi Marshes NR, Ceredigion – **Badger, Otter**
Tregaron Bog/Cors Caron NNR, Ceredigion – **Otter, Polecat**

Scotland

Abernethy Forest, Highland – **Pine Marten, Wild Cat**
Achnashellach, Highland – **Pine Marten**
Ardcastle, Argyll & Bute – **Pine Marten**
Ardnamurchan Peninsula, Argyll & Bute – **Otter, American Mink, Pine Marten, Wild Cat**
Beinn Eighe, Highland – **Pine Marten**
Culbin, Moray – **Badger**
Duncansby Head, Highland – **Wild Cat**
Glen Affric, Highland – **Stoat, Weasel, Pine Marten**
Glen Tamar, Grampian – **Wild Cat**
Glen Urquhart, Highland – **Stoat, Pine Marten**
Inner Hebrides and mainland Argyll & Bute – **Otter, American Mink**

Insh Marshes RSPB reserve, Highland – **Otter**
Knapdale, Argyll & Bute – **Pine Marten, Wild Cat**
Monaughty, Moray – **Stoat, Weasel, Pine Marten**
Outer Hebrides – **Otter, American Mink**
Plodda, Highland – **Pine Marten**
Ryvoan Pass near Aviemore, Highland – **Wild Cat**
Shetland – **Otter**

STATUS AND DISTRIBUTION

Red Fox
Population 258,000*
Range An extremely common mammal in Britain and Ireland. It is only absent from the Isles of Scilly and the Channel Islands.
Habitat Found in a diverse range of habitats from sand dunes and salt marshes to upland areas. Individuals have adapted very successfully to life in urban areas. Territories in urban areas can be as small as 0.2 km^2, while in hill country they may be as large as 40km^2.

Badger
Population 275,000
Range Widespread throughout Britain and Ireland. Scarce in northern Scotland. It is absent from higher areas (generally over 500m), lowland floodplains, intensively farmed agricultural land, some large cities, and most offshore islands. Densities can reach 38 adults per square kilometre in good habitat.
Habitat Common in lowland woodland and pastures but also occurs in some upland and urban habitats. It shows a marked preference for deciduous and mixed woodlands, but hedgerows and scrub, coniferous woodland, and open areas are also utilised. Setts are often located in sloping well-drained banks. It tends to select undisturbed areas with well-drained soils and good cover.

Otter
Population 12,900
Range Significant numbers occur in northern and western Scotland. Other substantial populations occur elsewhere in Scotland and in Wales and Ireland. Populations in England are showing signs of recovery in many areas, helped in certain river systems by reintroduction programmes. In some regions the Otter is now even being seen in urban areas.
Habitat Populations in coastal areas utilise shallow, inshore marine areas for feeding but also require fresh water for bathing, and terrestrial areas for resting and breeding holts. Inland populations utilise a range of running and standing fresh waters with adequate cover and food supplies.

Polecat
Population 63,200
Range Wales is still a stronghold, but it is spreading out across the English Midlands where it has now reached Cheshire, Northamptonshire and Oxfordshire, and it has been suggested that

there are now more in England than in Wales. Introductions have also seen this species breeding in Argyll, Cumbria, Hertfordshire, the Chilterns, and parts of Hampshire and Wiltshire. The Polecat now occurs as far east as Bedfordshire and Hertfordshire with smaller numbers in Essex, although it is not clear whether these represent animals spreading east from their native population or animals spreading out from the reintroductions into Hertfordshire in the 1970s.
Habitat The Polecat occurs in a wide range of habitats below 500 metres. It is found in wooded, coastal, wetland and riverine habitats and is also known to lie up in rat-infested farm buildings in winter.

Feral Ferret
Population 2,500, c.90% in Scotland
Range The Feral Ferret is known to occur commonly on Mull, Lewis, Bute, Arran, North and South Uist, Islay, Shetland and at several places on the mainland. It is possible that some 'Polecats' in eastern England may be Polecat/Feral Ferret hybrids.

American Mink
Population 36,950*
Range The Mink occurs throughout mainland Britain and on the Outer Hebrides where it poses a real threat to ground-nesting birds. A five-year project to eradicate Mink from North Uist and Benbecula was instigated in 2001.
Habitat The Mink is mainly found in aquatic habitats with adequate cover but may also spend a substantial time away from water even in urban areas if prey is abundant. Large populations also occur on rocky coasts, and around estuaries. In coastal areas it tends to forage in the mid-tide zone.

Stoat
Population 472,000*
Range The Stoat can be found at all altitudes throughout mainland Britain and Ireland. It also occurs on many larger offshore islands including mainland Shetland (where it was introduced in the 17th century or before), the Isle of Wight, Anglesey, the Isle of Man, Mull, Islay, Skye, Jura, Guernsey and Jersey.
Habitat The Stoat can be located in a variety of habitats throughout Britain and Ireland although favoured habitats include meadows and young forestry plantations. It uses features such as hedgerows and dry stonewalls as corridors between sites.

Weasel
Population 450,000
Range Common and widespread throughout Britain but absent from Ireland (where the Stoat is sometimes called the Weasel) and islands smaller than 380 square kilometres. It is present on Skye, Anglesey, Sheppey and the Isle of Wight.
Habitat The Weasel occurs in a wide range of different habitats with woodland, hedgerows and grassland being particularly preferred. Like the Stoat, it requires cover and linear features along which to travel.

Pine Marten
Population 3,650*
Range In Scotland, its stronghold is in the Highlands, but it may also be found in Dumfries and Galloway, Strathclyde, Grampian and Tayside. English refuges include Cumbria, Durham, Northumberland, and North and West Yorkshire, while in Wales it may still be found in small numbers in Clywd, Gwynedd, Powys and Dyfed. The position in north-east England is confused by the discovery of several American Martens in Kielder Forest, with another having been shot in Yorkshire in 1969. It is unclear whether the latter species is breeding in the area.
Habitat Inhabits Caledonian forests, young conifer plantations, coarse grassland, heather and grass moorland, or grass and scrub rides and borders. Although often associated with remote areas the Pine Marten frequently lives in close proximity to humans. It regularly raises young in occupied buildings, for example, on the Ardnamurchan Peninsula in western Scotland.

Wild Cat
Population 400–2,100 plus 1,200,000 Feral Cats
Range The Wild Cat is still reported in north-east Scotland, Easter Ross, north-east Inverness-shire, Strathspey, and parts of Perthshire and Argyll, but the purity of these individual populations is unclear. Population density appears to be highest in Deeside. More information can be found on www.scottishwildcats.co.uk
Habitat Usually found well away from humans in remote woodlands and moorland, although areas of pasture where prey abounds are also attractive to this species. It normally occurs below 600m.

* Excluding Ireland

SEALS

Common (Harbour) Seal *Phoca vitulina*
Grey Seal *Halichoerus grypus*

Two species of seal breed around the British coast – the Common (Harbour) Seal and the Grey Seal. Identification of both species is relatively straightforward with practice and there are readily accessible sites where both may be observed. Five additional species have occurred as vagrants and a Northern (Steller's) Sea Lion of unknown origin lived for several years on and around the Brisons, off St Just, Cornwall. It was last reported in 1998.

IDENTIFICATION

Common Seal Seals can be tricky to identify at sea where the head is often the only visible feature. There are differences in size between the Grey and Common Seals as well as spot patterns but the safe way is to concentrate on the head. The Common Seal usually shows numerous small spots on the head whilst the forehead is noticeably concave giving the animal a blunt-headed, almost dog-like appearance. It often basks in a characteristic manner with its body arched so that the hind flippers and the head are raised. The Grey Seal does not adopt this pose.

Grey Seal The forehead is straight to convex in the Grey Seal as opposed to concave in the Common Seal although in females and juveniles this is not so obvious. The muzzle is very long with the nostrils being almost parallel and not meeting at the base. Adult males are dark grey with light patches distributed liberally over the body. Females are a lot smaller and are generally silver-grey in colour with brown patches. The head shape of young Grey Seals is similar to that of Common Seals.

SEEING SEALS

Common Seals are easy to see, particularly during the moult from mid-August when up to 1,000 may haul out on the same sandbar. The pupping period between June and September is another excellent time to watch seals when they spend long periods ashore.

Male and female **Grey Seals** come ashore to moult at different times. Females moult from December through to March, whilst males moult from March until May. They may stay ashore then for several weeks and this is a good time to observe the species. The breeding season from September to November (peaking in October) is equally good, with seals spending long periods ashore. Studies have shown that between July and January some seals may haul out for up to five hours, spending about 35 hours foraging at sea in between.

KEY SITES

South-west England
Isles of Scilly – **Grey Seal**
Penwith Peninsula, Cornwall – **Grey Seal**

South-east England
Herne Bay, Kent – **Common Seal, Grey Seal**

East Anglia
Blakeney Point, Norfolk – **Common Seal, Grey Seal**
Horsey and Winterton-on-Sea, Norfolk – **Common Seal, Grey Seal**

Midlands
Donna Nook NNR, Lincolnshire – **Grey Seal**

Northern England
Seal Sands, Teeside – **Common Seal, Grey Seal**
The Farne Islands, Northumberland – **Grey Seal**

Wales
The Lleyn Peninsula and Bardsey Island, Gwynedd – **Grey Seal**
Strumble Head and the Pembrokeshire Coast – **Grey Seal**

Scotland
Ardnamurchan Peninsula, Argyll & Bute – **Common Seal, Grey Seal**
Dornoch Firth, Highland – **Common Seal, Grey Seal**
Firth of Tay and Eden Estuary, Angus/Fife/Perth & Kinross – **Common Seal**
Isle of May, Firth of Forth, Forth – **Grey Seal**
Inner Hebrides and mainland Argyll & Bute – **Common Seal, Grey Seal**
Isle of Skye, Highland – **Common Seal, Grey Seal**
Kinloch, Highland – **Common Seal, Grey Seal**
Knapdale, Argyll & Bute – **Common Seal**
Moray Firth and Cromarty Firth – **Common Seal, Grey Seal**
Orkney – **Grey Seal**
Outer Hebrides – **Grey Seal**
Shetland – **Common Seal, Grey Seal**

STATUS AND DISTRIBUTION

Common Seal
Population 40,000–65,000, 95% in Scotland
Range Locally common around the British Isles in both Shetland and Orkney, the Outer Hebrides, around the Scottish coastline, at several locations down the English east coast, particularly around the Wash and East Anglia and sparingly along the English south coast from Kent to Dorset.
Habitat A wide range of marine habitats including coastal lagoons, estuaries and rocky coasts. It is occasionally seen in rivers well away from the sea. It hauls out on sandbars, inter-tidal rocks and ledges, and on sandy and pebble beaches.

Grey Seal
Population 130,000, 90% in Scotland
Range Locally common in the British Isles. It occurs on all our coastlines breeding in most areas. It is also found on most of the offshore islands including Orkney, Shetland, Hebrides, the Isles of Scilly, the Isle of Wight, Anglesey and the Isle of Man.
Habitat The Grey Seal shows a clear preference for rocky coastlines but can also be found in sandy areas including estuaries and beaches.

VAGRANT SEALS

Walrus *Odobenus rosmarus*
Between 1815 and 1954 27 walruses were seen or killed in British waters. All, except one that was shot in the Severn in 1839 and one seen in the River Shannon in Ireland, were in Scotland. Since that time just eight have been recorded. In 1981 they were seen in Shetland, Arran and in the Wash area. In 1984 one was seen in the Pentland Firth and in 1986 two were in Shetland and one in Orkney. The most recent record is of a juvenile seen off Shetland in 2002.

Bearded Seal *Erignathus barbatus*
Bearded Seals appear off Britain more frequently than other vagrant seals. Shetland alone has recorded 12 individuals, the most recent of which occurred in May 2005. A long-staying individual spent several weeks in Hartlepool harbour in 1999 and another has spent a prolonged period in north Kent. The first Bearded Seal for Ireland appeared off County Mayo in 2002.

Harp Seal *Pagophilus groenlandicus*
There have been over 30 British records of this highly migratory species. Records extend right down the east coast from Shetland to Kent. On the west coast there have been records from Ayrshire down to Teignmouth in Devon with the most recent being an individual that turned up on the coast of Portland, Dorset, in September 2003.

Ringed Seal *Pusa hispida*
Relatively sedentary but occasional stragglers move south to Britain, for example, individuals off Norfolk in 1846, off Lincolnshire in 1889, Collieston, Grampian in 1897, Aberdeen in 1901, a juvenile on the Isle of Man in 1940, one shot on Whalsay in 1968 and one found on Orkney in September 1992. The most recent record was an individual that spent two days at Cullivoe, Yell, Shetland in July 2001.

Hooded Seal *Cystophora cristata*
Hooded Seals are very occasional stragglers to Britain with British records from Suffolk, Fife, Pembrokeshire, the Outer Hebrides, Cheshire, Orkney, Grampian and Shetland. The most recent sightings were in Shetland and Pembrokeshire in 2001.

CETACEANS

Humpback Whale *Megaptera novaeangliae*
Fin Whale *Balaenoptera physalus*
(Northern) Minke Whale *Balaenoptera acutorostrata*
Killer Whale *Orcinus orca*
Harbour Porpoise *Phocoena phocoena*
(Short-beaked) Common Dolphin *Delphinus delphis*
(Common) Bottle-nosed Dolphin *Tursiops truncatus*
Atlantic White-sided Dolphin *Lagenorhynchus acutus*
White-beaked Dolphin *Lagenorchynchus albirostris*
Long-finned Pilot Whale *Globicephala melas*
Risso's Dolphin *Grampus griseus*

At least 26 species of cetacean (whales and dolphins) are known to have occurred in British and Irish waters, and while many are very rare and only occur occasionally, at least a dozen are regularly seen around our coastlines. Unfortunately cetacean identification can be very difficult and many sightings may not be identifiable to species. Nevertheless, it is an often highly rewarding area of mammal watching.

IDENTIFICATION

Humpback Whale One of the easiest of the large baleen whales to identify. The stocky body with a generally low stubby dorsal fin on the rounded back are diagnostic particularly during dives. Other key features include the very long pectoral fins, the black and white markings on the undersides of the flukes and the knobs on the lower jaw. The blow is usually low, broad and bushy.

Fin Whale The Fin Whale's huge size is exceeded only by the Blue Whale. A sleek animal, it is dark grey above and white below. Other key identification features include its backward sloping dorsal fin, the single longitudinal ridge on the head and the very tall and narrow blow. The flukes are rarely raised when diving, when the blowholes surface shortly before the dorsal fin. A diagnostic feature usually noticeable at close range is the asymmetrical pigmentation of the lower jaw. The right lower jaw is white and the left side is black.

Minke Whale The Minke Whale should be readily identified by the trained eye. This small baleen whale is black or blue-grey above and white below. There is often a distinctive grey chevron across the back. The broad white band crossing the flippers is diagnostic. When surfacing the falcate (sickle-shaped) dorsal fin and blowhole appear simultaneously and during the dive, the flukes are never raised above the ocean surface. The blow is rarely visible.

Killer Whale The Killer Whale is not difficult to identify. The distinctive black and white body pattern combined with the tall dorsal fin – up to 1.8 metres high in mature males – and large size should be enough to secure a positive identification.

Harbour Porpoise The only porpoise species to be found in the seas of Britain and Ireland and its small size and robust shape are distinctive. However, although the average adult Bottlenose Dolphin is twice as long as a Harbour Porpoise, size illusion and light conditions can often create a misleading impression. Given this, the key identification feature is the distinctive surfacing roll that is easy to recognise once you are familiar with it. This rapid roll is accompanied by little or no splash as the animal exposes the small, triangular dorsal fin at the centre of a grey back. The flanks are paler while the belly is white, although the latter is rarely seen.

(Short-beaked) Common Dolphin Any large schools of dolphins seen in British and Irish waters are likely to be this species although a number of features need to be checked to confirm the identification. The dorsal fin, which is tall and falcate, and the upper back are dark grey or black forming a saddle with a downward point directly below the fin. Below this a yellow or tan patch to the front of the dorsal fin forms a distinctive hourglass pattern with the greyish flanks. The underside is white. This species may be confused with the superficially similar Atlantic White-sided Dolphin but is a much slimmer animal only being around half the weight of the latter species.

(Common) Bottle-nosed Dolphin This is a relatively robust dolphin with a short, but distinct beak. The dorsal fin is moderately tall and falcate. The body is generally unmarked with no sharp demarcation between the various grey shades.

Atlantic White-sided Dolphin The upper side is black or dark grey. It has a grey stripe along the flanks extending from the short beak to the flukes, a white flash between the flank stripe and the dark upper parts and a yellow flash behind the white flash. The underside is white. The only other British dolphin with a white flash below the dorsal fin is the White-beaked Dolphin but this is a much stockier species with considerably less contrast between the upper parts and the flank markings.

White-beaked Dolphin Stockier than the Atlantic White-sided Dolphin with a complex pattern of grey, black and white markings. The colours of a White-beaked Dolphin are less clearly demarcated than those of a White-sided Dolphin and are highly variable. The short thick beak is usually white but may be grey or even brown. The prominent falcate dorsal fin is dark grey with the coloration extending onto the back to create a saddled effect. There is a second greyish-white saddle behind the dorsal fin. The coloration below the saddles is diffuse with a combination of broad greyish-white blazes on the side and generally a distinct dark grey patch in the middle of the flanks.

Long-finned Pilot Whale A stocky, medium-sized cetacean with a bulbous head, distinctive lobed dorsal fin and long, sickle-shaped flippers. Generally black in coloration, often with a pale saddle. Very sociable, often associating with smaller cetaceans.

Risso's Dolphin One of the easiest dolphin species to identify. Its large size, large rounded (almost blunt) head, tall dorsal fin, and extensive scarring on the body are all distinctive. Older animals may be almost white.

SEEING CETACEANS

Few cetaceans are regularly seen from the mainland and it will be necessary to go to sea to observe many of the species. Fortunately there are now numerous private companies offering boat trips to see cetaceans, particularly off the western coast of Scotland and the southwest coasts of Ireland, during the summer months. In addition regular ferry crossings, particularly those between England/Wales and Ireland, and those between the Scottish mainland and the western and northern isles, offer excellent opportunities to see cetaceans. Ferries from Portsmouth and Plymouth to the northern coast of Spain also offer great opportunities for seeing cetaceans. However, the majority of those sightings tend to be in the Bay of Biscay and few cetaceans are actually seen in British waters.

Humpback Whales are annual but rare in British waters: e.g. between 1975 and 1991 there were 13 sightings in Scottish waters. In recent years, probably as a result in the increased interest in watching cetaceans, the species appears to have become commoner. There is a small chance of observing this species at any time of year although in recent times the most regular Humpbacks have started to turn up in the winter months. Long-staying individuals reported on bird information services offer the best opportunity for seeing this species in Britain.

Fin Whales are regularly seen from land in some areas, and prominent headlands overlooking deep water are the best areas to try. They appear to be most common off British coasts from April to December although there were several sightings off the south coast of Cornwall in January and February 2005 and no fewer than 83 individuals were seen off southern Ireland in December 2005 and January 2006. Boat-based searches can be rewarding as they appear indifferent to boats and often allow close approaches. Recent observations of Fin Whales from Britain/Ireland ferries suggest that the Irish Sea is a popular feeding area or migratory route for this species. Routes between Pembroke and Rosslare and between Swansea and Cork have recently reported animals. Summering animals have become increasingly common off southwest Ireland in recent years and several operators offer trips out to look for these animals.

Minke Whales are common in the waters off western Scotland from June to October and boat-based searches such as those off Mull and the Ardnamurchan Peninsula are likely to be successful throughout this period. September and October can be particularly rewarding as the whales actively feed in preparation for their migration. Groups of feeding seabirds, particularly juvenile gulls, Gannets or shearwaters, are often indicators of rich feeding areas. Such areas are equally likely to attract feeding Minke Whales and other cetaceans and locating congregations of feeding seabirds is often a good way of finding cetaceans. Seawatching from Scottish headlands, inter-island ferries and dedicated cetacean watching trips all offer opportunities to see this species.

Killer Whales are mainly recorded between April and October, although they have been reported in every calendar month. In the summer months pods of Killer Whales may linger around seal colonies in Shetland for days at a time, preying on both adults and young. This can provide opportunities not only to catch up with this species but also to see it hunting. Alternatively, as with most of the commoner cetaceans covered by this book, any ferry journey within the British Isles provides a chance of spotting a pod of Killer Whales. Ferries to and from the Outer Hebrides, and inter-island ferries within Shetland provide regular opportunities and ferries across the Irish Sea, e.g. from Pembroke to Rosslare occasionally turn up this species.

Harbour Porpoises occur in British waters throughout the year but peak numbers occur in March to April and July to November. Juveniles can be seen anytime between May and August, with a peak in June. They tend to occur in waters less than 100 metres deep. Harbour Porpoises are regularly seen feeding on the same shoals of fish as seabirds such as Gannets and Kittiwakes so feeding groups of these species are worth locating as they may lead you to cetaceans such as porpoises. Given their small size and inconspicuous dorsal fins Harbour Porpoises can be particularly difficult to spot in anything other than calm conditions. If sea conditions are rough, look for calmer inlets as they are known to search out calmer waters in inclement weather.

(Short-beaked) Common Dolphins are recorded in British and Irish waters throughout the year. Large numbers move into western Britain in winter when they follow mackerel inshore. They also occur off northern Scotland in August, when herring congregate to spawn, and off Ireland and south-western England in summer, when they are attracted by the strong movements of sprats, sand eels and mackerel.

(Common) Bottle-nosed Dolphins are generally found inshore and consequently can often be seen from land. In some of the sites listed the populations occupy large territories and you need to be lucky to be in the right place at the right time. This is the most frequently encountered dolphin species around our coasts and in some areas resident or semi-resident populations can readily be seen at certain times of the year. In the Moray Firth and Cardigan Bay several companies offer trips to watch the dolphins.

Atlantic White-sided Dolphins are difficult to see off Britain and Ireland. The northern archipelagos, including the Outer Hebrides, offer the best chance particularly on ferry crossings between June and November. In recent years there have also been regular records off the Aberdeenshire coast in midsummer. Aggregations of this species may number several hundred but rarely exceed 30 off Britain and Ireland. They are often found in the company of other cetacean species, in particular White-beaked Dolphins.

White-beaked Dolphins are one of the easier cetaceans to see from land: particularly in areas such as the Shetlands and the coast of Aberdeenshire. Inter-island ferries between the Scottish mainland and outlying island groups also offer good opportunities to see them. They occur in British waters throughout the year but are particularly common from June to October. Most are seen around the mainland and islands of northern Scotland.

Long-finned Pilot Whales are usually found in deep offshore waters although they are frequently encountered in coastal waters in certain areas of the North Atlantic. They appear to be common and widespread in deep northern European waters, but seasonally enter coastal areas around the Faeroe Islands, north Scotland, western Ireland and the Channel Approaches west of England. There is evidence that the species increased substantially in the region in 1970s and early 1980s but has declined since. Although they are mainly pelagic, they are seen in all coastal areas except the southernmost part of the North Sea. They have been recorded in every month of the year, but most records are from June to September in the north, and November to January further south. Long-finned Pilot Whales are becoming increasingly rare off Wales. During the 1970s and 1980s, this species accounted for over 2% of all cetacean sightings in this area. This decreased to 0.3% during the 1990s. In southwest England they remain rare but may be seen in any month of the year. The majority of sightings come from November, December and January. It has been suggested that these whales move in from the northern Atlantic in winter to feed on squid and on a range of commercial fish, particularly mackerel.

Risso's Dolphins are most abundant in British waters from May to October but they are being encountered increasingly in winter. Many of the sites detailed below are either archipelagos or offshore islands and while the chances of successful observations increase from a boat or further offshore they do occur near the coastline and can be viewed from prominent headlands in suitable locations.

KEY SITES

South-west England

Durlston Head and Country Park, Dorset – **Bottlenose Dolphin**
Isles of Scilly – **Common Dolphin**
Penwith Peninsula, Cornwall – **Minke and Killer Whales; Harbour Porpoise; Common, Bottlenose and Risso's Dolphins**
Portland, Dorset – **Bottlenose Dolphin**
Rame Head, Cornwall – **Common Dolphin**
Start Point and Berry Head to Sharkham Point NNR, Devon – **Harbour Porpoise**

East Anglia

Horsey and Winterton-on-Sea, Norfolk – **Harbour Porpoise**

Northern England

Whitley Bay, Northumberland – **Harbour Porpoise**

Wales

Cardigan Bay, Ceredigion – **Bottlenose Dolphin**
Strumble Head, Pembrokeshire Coast – **Killer Whale, Harbour Porpoise, Risso's Dolphin**
The Lleyn Peninsula and Bardsey Island, Gwynedd – **Harbour Porpoise**

Worm's Head and Burry Holms, Gower Peninsula, Glamorgan – **Harbour Porpoise**

Scotland

Ardnamurchan Peninsula, Argyll – **Minke Whale**

Eastern Scotland – **Minke Whale; Atlantic White-sided and White-beaked Dolphins**

Firth of Forth – **Humpback and Minke Whales, Harbour Porpoise**

Inner Hebrides and mainland Argyll & Bute – **Minke and Killer Whales; Harbour Porpoise; Common, Bottlenose and Risso's Dolphins**

Isle of Skye, Highland – **Minke and Killer Whales, Atlantic White-sided and White-beaked Dolphins**

Moray Firth and Cromarty Firth – **Minke and Killer Whales, Bottlenose Dolphin**

Outer Hebrides – **Killer Whale; Atlantic White-sided, White-beaked and Risso's Dolphins**

Shetland – **Humpback, Fin, Minke and Killer Whales, Harbour Porpoise; Atlantic White-sided and White-beaked Dolphins**

Ireland

Humpback, Fin, Minke and Killer Whales; Harbour Porpoise; Common, Bottlenose, Atlantic White-sided and Risso's Dolphins

STATUS AND DISTRIBUTION

Few accurate estimates are available for the populations of cetaceans around the British and Irish coast.

Humpback Whale

Range Seen annually off Scotland and Ireland with regular records off Shetland.

Fin Whale

Range Fin Whales are seen regularly from land particularly from viewpoints overlooking deep water. South-west Ireland offers the best opportunity for encountering the species. It appears to be most common from April to December.

Minke Whale

Range Common off western Scotland and Shetland in summer. Frequent off eastern Scotland, Ireland and Wales; scarce elsewhere.

Killer Whale

Range Occurs most frequently around Shetland and the Outer Hebrides. Scarcer elsewhere.

Harbour Porpoise

Range Restricted to coastal waters, it frequents bays, estuaries and harbours. The majority of animals stay within 10km of land. Occurs throughout the British Isles but commonest in Scotland. It has declined over the last four decades along the southern North Sea and Channel coasts due in part to accidental drowning in fishing nets, pollution of coastal waters, and the changing dynamics of prey stocks (particularly herring). The British population, i.e. those in waters up to 200 nautical miles offshore, has been estimated at 120,000 (Hammond *et al.*, 1995).

(Short-beaked) Common Dolphin
Range One of the most widely distributed cetaceans in the world and one of the most frequently sighted in the south-west of Britain and Ireland. It is thought that it has also been increasing in numbers off north-west Scotland and in the northern North Sea since the mid-1990s. Large numbers move into the western English Channel during the winter with a southerly movement in late spring.

(Common) Bottle-nosed Dolphin
Range Occurs widely but may be declining around much of our coastline. There are resident groups of at least 130 animals in Cardigan Bay in Wales and in the Moray Firth in Scotland, and another 40 occur off the coast from Cornwall to Dorset. It is also resident off the west and south-east coasts of Ireland while transient groups occur widely elsewhere. The Moray Firth population appears to be extending its range and has been seen as far south as Fife. This species is extremely adaptable and can be found in a variety of habitats throughout its range. Around the British Isles, typical habitats include bays, estuaries and inshore waters centred on islands. It will enter river systems with individuals occasionally being seen well inland.

Atlantic White-sided Dolphin
Range Generally considered to be a deep-water species. It is believed to be relatively common along the continental shelf west of the Atlantic coast of Ireland and Scotland, and in the North Sea. Groups of 6–30 individuals have been seen in coastal waters with larger groups further offshore.

White-beaked Dolphin
Range Favours continental shelf areas and appears to be more common in coastal areas than the Atlantic White-sided Dolphin. The species is common in the North Sea and around the northern archipelagos. Formerly common off the west coast of Scotland, recent research suggests that it is being replaced by the Common Dolphin possibly as a consequence of higher sea temperatures.

Long-finned Pilot Whale
Range Occasionally encountered in coastal waters around north Scotland, western Ireland and south-west England. Most records are in June to September in the north, and November to January further south. Rare in south-west England but may be seen in any month.

Risso's Dolphin
Range Widely distributed, occurring in all the world's tropical and temperate seas. Although taken occasionally as fishing by-catch, this is still an abundant species although there are no recent estimates of the population around the British and Irish coasts. It is rare in the southern part of the North Sea but not uncommon around the Outer Hebrides and in the Irish Sea.

RARE AND VAGRANT CETACEANS

Vagrant species of whale and dolphin occasionally turn up off our coastlines. Those below have either stranded on beaches or been identified at sea.

Beluga *Delphinapterus leucas*
Belugas have a circumpolar distribution in the Northern Hemisphere between 50°N and 80°N. There were seven sightings and a single stranding (in County Mayo) prior to 1991. The most recent record of this rare vagrant is one off Shetland in 1997. Surprisingly, another well-watched individual fed in a sea-loch in the same archipelago in 1996.

Narwhal *Monodon monoceros*
The Narwhal is another high Arctic species with six British records up until 2001. Five of the records relate to strandings. The other was a sighting of two off Orkney in 1949.

North Atlantic Right Whale *Eubalaena glacialis*
This species is now very rare in European waters, where it was once found from Norway and Iceland to the British Isles, France and Spain. There have been no strandings since 1913, but there were well-documented sightings off Southern Ireland in 1970, in the northern Irish Sea in 1979 and off the Outer Hebrides in August 1980.

Blue Whale *Balaenoptera musculus*
The Blue Whale occurs in all the world's oceans and inhabits coastal, shelf, and oceanic waters. There have been no strandings in the region since 1930 although one was reported off the northwest coast of Ireland in May 1977. Recent summer sightings in the Bay of Biscay and records of a small winter population to the west of Ireland provide some hope that the species may be seen in British and Irish waters again.

Sei Whale *Balaenoptera borealis*
Sei Whales have a cosmopolitan distribution occurring from tropical waters to high latitudes and undertake seasonal movements between high latitudes in summer and tropical waters in winter. There were nine strandings between 1913 and 1987 and six sightings between 1975 and 1991 although three of these were well offshore from weather ships. They are rarely seen close to shore but they have recently been observed, along with Fin Whales, migrating past headlands in County Cork, Ireland, mainly in late autumn/early winter. Headlands that may be worth investigating there include the Old Head of Kinsale, Galley Head and Glandore. Sei Whales are also regular in the Faeroe-Shetland Channel.

(Great) Sperm Whale *Physeter macrocephalus*
Sperm Whales inhabit ice-free deep water from the equator to the edge of the polar pack ice. Within Britain sightings and strandings of this species are generally restricted to the northern isles and the Hebrides. The Fair Isle Channel provides deep-water conditions (over 1,000m deep), and the Shetland Sea Mammal Report 1998 notes a group of 12–14 individuals in July that year. The 2002 report details a sighting of a subadult in October that year from the M.V. *Good Shepherd*. A male was also observed in Weisdale Voe in January 1999. Strandings appear to have increased since

1900 and this species regularly strands in Shetland and along the east coast. There have been at least three strandings on the Norfolk coast in the past 25 years, an adult spent at least a week in the Firth of Forth in 1997 and another was stranded in the Humber Estuary in February 2006.

Pygmy Sperm Whale *Kogia breviceps*
Occurs in deep water in tropical, subtropical and warm temperate waters. There were strandings in County Clare in April 1966 and south Wales in October 1980. There have also been three sightings of whales thought to be this species: one in August 1979 in the North Sea, and another on two successive days in June 1982 in deep waters off north-west Ireland.

Northern Bottle-nosed Whale *Hyperoodon ampullatus*
Northern Bottle-nosed Whales are found in both Arctic and cold-temperate waters. They favour deep waters over submarine canyons and along the edge of the continental shelf. Most sightings off Britain occur in summer and autumn as the species migrates south. They are frequently seen 500–700km west of Scotland: at least 215 sightings from 1960 to 1985, and there were over 100 strandings between 1913 and 1991, but coastal records are extremely infrequent. Most of the reports received are from the western coast of Scotland. In 1998 a pair of Northern Bottle-nosed Whales stayed in Broadford Bay on the Isle of Skye for three weeks. A second pair was found in a similar shallow-water loch on the Isle of Bute at the same time. Other recent records include individuals off the coast of north-west England, particularly Cumbria. The most recent records involved one off Dursey Head, County Cork, in August 2004 and the famous individual that entered the River Thames and reached as far upstream as Central London in January 2006 (with another being seen off the Essex coast at the same time).

Cuvier's Beaked Whale *Ziphius cavirostris*
Cuvier's Beaked Whale has the widest distribution of any beaked whale species. In the Atlantic, strandings have occurred as far north as the Shetland Isles and as far south as southern Africa. There were three documented sightings in the 1980s (two in Ireland) and another two were seen six miles south of Cape Clear Island in June 1998. There have been at least 40 strandings since 1913 including one in east Norfolk in June 2002.

True's Beaked Whale *Mesoplodon mirus*
True's Beaked Whale occurs throughout the world's oceans. Of the nine European strandings up to 1991 six are from the west coast of Ireland with another in the Outer Hebrides.

Gervais's Beaked Whale *Mesoplodon europaeus*
This is an Atlantic species, which has never been positively identified at sea. It appears to be found only in the tropical and warm temperate areas of the Atlantic. There have been recent strandings in Ireland.

Sowerby's Beaked Whale *Mesoplodon bidens*
Sowerby's Beaked Whale occurs in cool and warm temperate waters in the North Atlantic. Individuals frequently strand in Britain and Ireland with at least 43 stranding between 1913 and 1985. Two different individuals stranded within days of each other along the coast of south Wales in

September 2004, one stranded in County Wexford in August 2004 and another in Aberdeenshire in January 2005. The only documented sighting relates to one seen off the west coast of Scotland in August 1977.

Blainville's Beaked Whale *Mesoplodon densirostris*
This is a widespread species, found in tropical and warm temperate waters in all of the world's oceans. Strandings have occurred in Britain and Ireland.

False Killer Whale *Pseudorca crassidens*
A widely distributed species in tropical, subtropical and warm temperate waters worldwide. There were mass strandings in Scotland (c.150) in 1927, and in South Wales in 1934 (c.25) and 1935 (c.75). Between 1976 and 1991 there were four sightings between 5km and 54km offshore including a group of over 100 off north-east Scotland. Since then five were seen together off Black Head, County Clare, in May 1995 and more recently, two were seen together off the southwest of Ireland in November 2000.

Striped Dolphin *Stenella coeruleoalba*
The Striped Dolphin is a cosmopolitan species, occurring in subtropical and warm temperate waters worldwide. There is a large population in the Bay of Biscay in summer, yet it remains rare in British and Irish waters with around 70 animals being sighted or stranded in the 1990s. Striped Dolphins are occasionally seen from the Isles of Scilly, and from ferries operating between Britain and Ireland and off south-west Wales. There are few records from the well-studied waters 500–700km west of Scotland suggesting that their northward distribution may be governed by the Gulf Stream.

Fraser's Dolphin *Lagenodelphis hosei*
Fraser's Dolphin occurs primarily in deep tropical waters. In the Atlantic they generally do not occur north of Morocco. One was found beached on South Uist, Western Isles, on 3 September 1996.

UNGULATES

(Eurasian) Wild Boar *Sus scrofa*
Red Deer *Cervus elaphus*
Sika Deer *Cervus nippon*
Fallow Deer *Dama dama*
(Western) Roe Deer *Capreolus capreolus*
(Reeves's) Muntjac *Muntiacus reevesi*
Chinese Water Deer *Hydropotes inermis*
Feral Goat *Capra hircus*

Excluding **Reindeer** (Caribou), **Feral Cattle, Wild Horses** and most breeds of **Feral Sheep**, which only survive in Britain in a managed state, there are eight species of ungulate which occur in Britain in a feral or wild state. **Père David's Deer** is kept in a number of deer parks, and escapes have taken place with some animals existing in the wild for a short time but no feral populations have become established. In addition **Soay Sheep**, perhaps the most primitive extant form of domestic sheep, exist in unmanaged populations on the islands of Hirta and Soay in the St Kilda archipelago, and have done so for perhaps 4,000 years.

IDENTIFICATION

Wild Boar and Feral Goats are easily identified.
Identification of deer should generally pose no problems although individual animals in their drab winter coats can be problematical. It is important to concentrate on two key features – the rump and tail pattern and the shape and style of the antlers (if applicable).

Wild Boar Only likely to be confused with feral pigs but readily identified by the bristly brown coat, which turns grey with age, its long pointed head and distinctive upper canines. Some animals in Kent/East Sussex are very pale, suggesting interbreeding with pigs.

Red Deer A large deer standing about 1.2m at the shoulder, with a reddish-brown coat during summer and a greyish-brown coat in winter. The cream-coloured rump lacks the black border of the Sika and Fallow Deer. The antlers on mature stags are well branched with many points. Red × Sika hybrids tend to have some spotting.

Sika Deer Pure Sika Deer are intermediate in size between Red and Roe Deer. The summer coat is reddish with distinct white spots. The winter coat is grey to almost black with indiscernible spots. The tail and rump are white and the rump has a dark upper border. A pale oval-shaped mark on the hocks is diagnostic.

Fallow Deer Fallow Deer occur in a variety of colour morphs ranging from black, through shades of brown with white spots, to all white. The winter coat, which is much thicker, is usually a lot darker. Has a white rump edged with black and a black line down the long tail. This is the only British deer species with palmate (broad-bladed) antlers.

Roe Deer A medium-sized deer standing 0.6m at the shoulder. Brown with a very short tail and a very pale buff rump in summer, greyer with a white rump in winter. The male has short, stubby antlers. The face is distinct with a white chin contrasting strongly with the black nose and black eyes.

(Reeves's) Muntjac A very small reddish-brown deer with distinctive facial markings: black stripes forming a V, and a white underside to the tail that is conspicuous when the animal is alarmed. The male has short antlers. Runs with the head held low and the hindquarters raised giving a characteristic hunched appearance not unlike a large dog.

Chinese Water Deer Reddish-brown in summer, sandier in winter. Slightly taller than the Muntjac with a straight (not hunched) back and a short tail. No facial markings or antlers but the male has protruding tusks and both sexes have wide rounded ears.

Feral Goat Highly variable in colour but only likely to be confused with primitive species of sheep. The beard and the callus on the knees are diagnostic.

SEEING UNGULATES

Wild Boar are mainly nocturnal emerging to feed along forest rides and in adjacent fields at dusk. Damage to fencing, areas of digging, notches on trees and sausage-shaped faecal pellets up to 7cm thick and 10cm long are all indications that there are Wild Boar in the area. Boar can be aggressive so females with young should never be approached.

Red Deer are fairly easy to observe especially in open habitats. They are most active around dusk and dawn and are particularly visible during the rut between August and October.

Sika Deer are generally easy to see in open areas such as fields and woodland edges, particularly during the rut from late August to October. However, like most deer, they are fairly shy and long-range views are the norm.

Fallow Deer are easy to see in many areas, particularly towards dusk, by walking woodland edges and adjacent farmland. Alternatively sitting and waiting in suitable habitat at dusk is also likely to lead to success.

Roe Deer can be seen throughout the day but are most conspicuous at dusk and dawn. Walking down woodland tracks or sitting quietly in suitable-looking habitat at dusk may produce sightings. They can be very vocal and the bark of the buck is often the first indication that there are Roe Deer nearby.

Muntjacs are easily seen throughout southern Britain, i.e. south of a line from the Wirral to the Humber, even in the middle of towns. They can be seen throughout the day and can be confiding in areas where they have frequent contact with people. They are very vocal and the barking of the male, which gives the species its alternative name 'Barking Deer', is a common sound particularly at dusk.

Chinese Water Deer frequently emerge from woodlands and marshes to feed in open fields at dusk. They are particularly conspicuous during the rutting season, November to January, when the bucks can be very vocal.

Feral Goats can be remarkably difficult to find and a lot of effort may be necessary scanning hillsides and cliffs where they occur. They may be easier to locate on islands such as Lundy, where their habitat is more limited.

KEY SITES

South-west England

Arne RSPB Reserve, Dorset – **Sika Deer, Roe Deer**
Brean Down, Somerset – **Feral Goat**
Brownsea Island, Dorset – **Sika Deer, Fallow Deer**
Cardinham Woods, Cornwall – **Red Deer, Roe Deer**
Durlston Head Country Park, Dorset – **Roe Deer**
Exmoor National Park, Devon/Somerset – **Red Deer**
Haldon Woods, Devon – **Fallow Deer**
Hooke and Higher Kingcombe, Dorset – **Wild Boar, Fallow Deer, Roe Deer.**
Lydford, Devon – **Red Deer, Roe Deer**
Lundy, Devon – **Sika Deer, Feral Goat**
Savernake Tunnel, Wiltshire – **Fallow Deer**
The Quantocks, Somerset – **Red Deer**
Valley of the Rocks, Devon – **Feral Goat**
Wareham Forest and Ringwood Forest, Dorset – **Roe Deer**

South-east England

Bonchurch Down, Isle of Wight – **Feral Goat**
Bowdown Woods, Berkshire – **Roe Deer, Muntjac**
East Beckley Woods, Kent/East Sussex – **Wild Boar**
Micheldever Wood, Hampshire – **Fallow Deer, Roe Deer**
New Forest, Hampshire – **Red Deer, Sika Deer, Fallow Deer, Muntjac**

East Anglia

Bedford Purlieus NNR, Cambridgeshire – **Sika Deer**
Bradfield Woods NNR, Suffolk – **Roe Deer, Muntjac**
Breckland/Thetford Forest, Norfolk/Suffolk – **Red Deer, Roe Deer, Muntjac**
Hickling Broad and Stubb Mill, Norfolk – **Chinese Water Deer**
Holkham, Norfolk – **Fallow Deer, Muntjac**
Minsmere RSPB Reserve, Suffolk – **Red Deer, Chinese Water Deer, Muntjac**
Strumpshaw Fen RSPB Reserve, Norfolk – **Chinese Water Deer**
Woodwalton Fen, Holme Fen and Monk's Wood, Cambridgeshire – **Chinese Water Deer**

Midlands

Ashridge Park, Hertfordshire/Buckinghamshire – **Fallow Deer**
Bourne Woods, Lincolnshire – **Fallow Deer**
Broxbourne Woods, Hertfordshire – **Muntjac**
Finshade Wood (Rockingham Forest), Northamptonshire – **Sika Deer, Roe Deer, Muntjac**
Wardley Wood, Leicestershire – **Muntjac**
Wendover Woods, Buckinghamshire – **Muntjac**
Wye Valley and Forest of Dean, Monmouthshire/Gloucestershire – **Fallow Deer**
Wyre Forest, Worcestershire/Shropshire – **Fallow Deer, Muntjac**

Northern England

Boltby Forest, Yorkshire – **Fallow Deer, Roe Deer**
Forest of Bowland, Lancashire – **Sika Deer**
Grizedale Forest Park, Cumbria – **Red Deer**
Kielder Forest, Northumberland – **Roe Deer**
Leighton Moss RSPB Reserve, Lancashire – **Red Deer**
North Riding Forest Park, Yorkshire – **Roe Deer**
Wykeham Forest, Yorkshire – **Roe Deer**

Wales

Afan Forest Park, Neath – **Fallow Deer**
Brecon Beacons NP – **Red Deer**
Coed y Brenin, Gwynedd – **Fallow Deer**
Crychan, Powys – **Roe Deer**
Dyfi, Gwynedd – **Fallow Deer**
Nercwys, Denbighshire – **Muntjac**
Radnor Woods, Powys – **Roe Deer**
Snowdonia – **Feral Goat**

Scotland

Abernethy Forest, Highland – **Red Deer**
Achnashellach, Highland – **Red Deer, Sika Deer, Roe Deer**
Ardnamurchan Peninsula, Argyll & Bute – **Red Deer, Roe Deer**
Beinn Eighe, Highland – **Red Deer**
Cairnsmore of Fleet, Dumfries and Galloway – **Feral Goat**
Clashindarroch, Aberdeenshire – **Red Deer**
Culbin, Moray – **Roe Deer**
Findhorn Valley, Highland – **Red Deer, Sika Deer, Feral Goat**
Glen Tanar, Grampian – **Red Deer**
Glen Affric, Highland – **Red Deer, Sika Deer**
Inner Hebrides and mainland Argyll & Bute – **Feral Goat**
Kilmun Arboretum, Argyll & Bute – **Roe Deer**
Kinloch, Highland – **Roe Deer**
Mabie Forest, Dumfries and Galloway – **Roe Deer**
Palacerigg Country Park, Lanarkshire – **Roe Deer**
The Durris Forest, Aberdeenshire – **Roe Deer**

STATUS AND DISTRIBUTION

Wild Boar
Population <1,000
Range Populations are now established in Kent/East Sussex with smaller numbers in Dorset and the Forest of Dean.
Habitat Thick woodland and scrubby areas with thick ground cover. Will feed in adjacent farmland.

Red Deer
Population 316,000*
Range Common in the Scottish Highlands, Exmoor and the Lake District. Smaller numbers occur elsewhere in Scotland, in East Anglia, Hampshire/Dorset, the Midlands and Yorkshire.
Habitat Moorland in Scotland, Ireland and southwest England. Forested areas elsewhere but will also feed in adjacent farmland and marshes.

Sika Deer
Population 26,600*
Range Large populations in Argyll, Inverness, Sutherland, Peebleshire, Ross and Cromarty, Dorset, Hampshire, Northamptonshire and Lancashire.
Habitat Dense woodlands and scrubby vegetation along with young coniferous plantations. Feed in adjacent open fields.

Fallow Deer
Population 128,000*
Range Widespread but patchily distributed in mainland Britain.
Habitat Mature woodland, visiting adjacent open fields particularly from dusk to dawn.

Roe Deer
Population 300,000
Range Widespread in Scotland, the Lake District and southern and eastern England. Small numbers in Wales and on the Scottish islands of Bute, Skye, Islay and Seil.
Habitat Woodlands with thick cover where they can feed undisturbed. In some areas, e.g. the Brecks of East Anglia, also found feeding in open fields even in the middle of the day.

(Reeves's) Muntjac
Population 128,500
Range Widely distributed in England although commonest south of a line from the Wirral to the Humber, with smaller numbers in Scotland and Wales.
Habitat Woodland and low shrubby areas including parks and gardens in built-up areas.

Chinese Water Deer
Population 1,500
Range Isolated populations in Bedfordshire, Buckinghamshire, Cambridgeshire, Norfolk, Suffolk, Avon and possibly Hertfordshire.
Habitat Woodland and adjacent farmland in Bedfordshire and Buckinghamshire. In marshland and adjacent fields elsewhere.

Feral Goat
Population 3,565*
Range Scottish Highlands, Snowdonia, many parts of Ireland, the Isle of Wight, and a number of small islands off the west coast of Britain including Lundy.
Habitat Mainland animals occupy mountainsides in summer and adjacent valleys in winter. Island animals are often found on sea cliffs.

* Excluding Ireland

MARSUPIALS

Red-necked Wallaby *Macropus rufogriseus*

Originating from Australia, Red-necked Wallabies have become at least temporarily established in various parts of the British Isles. In Britain, Red-necked Wallabies inhabit scrub and woodland, requiring areas of low bush for cover from which they appear to feed in the open at dusk. Those introduced to Inchconnachan Island survive in an area of Caledonian Pines.

SEEING WALLABIES

In Australia and New Zealand Red-necked Wallabies emerge around dusk and continue to feed throughout the night. In Britain, however, particularly on Inchconnachan Island, a lack of predators has resulted in the wallabies remaining active throughout the day. Although the lack of cover may make viewing easier in winter, 'British' wallabies can readily be seen throughout the year and people have seen up to seven wallabies in less than a couple of hours during the summer months. They are shy by nature, however, and stealth and patience may be required to obtain good views.

KEY SITES

Northern England
Ballaugh Curraghs, Isle of Man

Scotland
Loch Lomond and Trossachs National Park, Clyde

STATUS AND DISTRIBUTION

The most famous population, in the Peak District, became established during the early 1940s and built up to about 50 animals. Bad winters, however, took their toll and this, coupled with disturbance at the site, led to a dramatic decline in the population. As at June 2005 only one animal survives. A second colony lived in Ashdown Forest in Sussex from about 1940 to 1972 while another small colony in Devon is believed to have died out in 1994. The third and currently most successful colony inhabits the Loch Lomond area where two pairs were released on Inchconnachan Island in 1975. The latest estimate is of at least 41 individuals. A fourth 'wild' population occurs at Ballaugh Curraghs on the Isle of Man. Originating from four individuals that escaped from a local wildlife park, the population appears to have grown to at least 50 individuals. Some reports suggest that there may be as many as 60 in the area. Finally, a free-ranging colony still occurs in the grounds of Whipsnade Park Zoo in Bedfordshire and another smaller colony may have become established nearby at an undisclosed location in Buckinghamshire.

INTRODUCTION TO THE SITE GUIDE

The sites listed here are by no means a full list of the best places to see mammals in Britain and Ireland. They are simply a selection of sites that offer good mammal-watching opportunities. Others, particularly many of the excellent reserves run by the RSPB or the network of county wildlife trusts, also hold a variety of mammals. Full details of these sites can be found on their websites (see Useful Addresses). Your comments or suggestions are very much welcomed and will be acknowledged in any future edition of this book. They may be sent directly to the author at rd_moores@yahoo.co.uk or to the publishers.

Grid reference – a fairly obvious starting point for each site account is the grid reference. This will allow you to pinpoint either the exact location (e.g. a car park, visitor centre or other definable location) or the central point of a reserve where each species occurs.

Key species – this section only highlights the key species at each site, but other more common species are also likely to be present. ***It must be remembered that species such as Hazel Dormouse and Harvest Mouse are extremely difficult to see even at the key sites, and that bats such as Whiskered and Brandt's are virtually unidentifiable in the field even with bat detectors.***

Access – this gives directions to each site as well as further information regarding opening times and facilities for visitors.

Site description – this details the habitat(s), the best areas and the times of day or year to look for the key species, and also includes other mammals that are present at the site. Such is the nature of many mammals that it is extremely difficult to pinpoint exact sites where it is possible to go and see a particular species. While I would like to be able to tell the reader to 'just go through the National Trust kissing gate, proceed for a further 28 metres and look left down the woodland ride and you should see a Pine Marten', this is just not possible. Consequently the information in this section varies greatly according to the target species, and the information simply provides guidance on areas where they occur and suggestions on where to look within these areas.

Other information – this section is included where relevant to provide further information of interest such as notable flora and other fauna at the site, and anything else that may be useful when visiting the site.

SOUTH-WEST ENGLAND

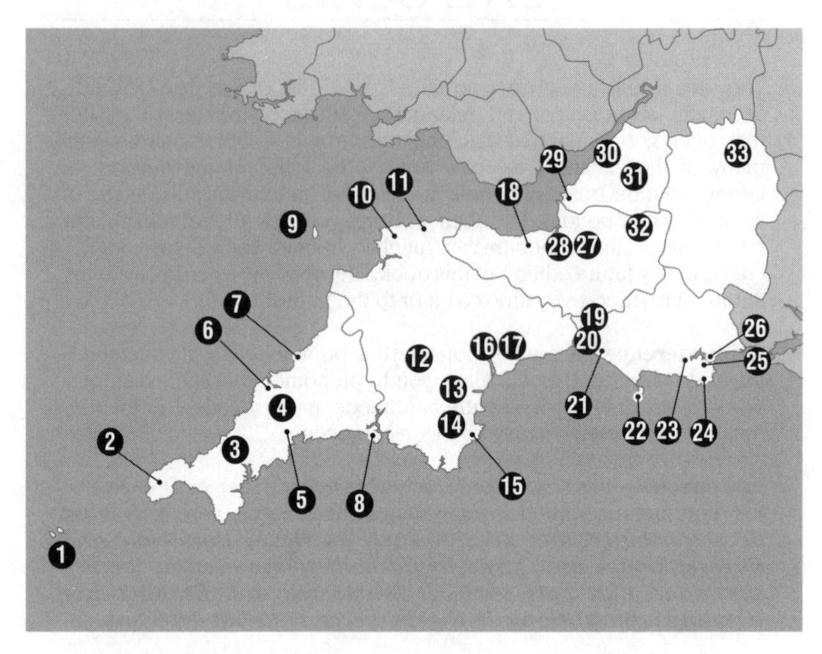

1 Isles of Scilly
2 Penwith Peninsula
3 Five Acres CWT Reserve
4 Cardinham Woods
5 Prideaux Wood
6 River Camel
7 Minster Church SSSI
8 Rame Head
9 Lundy
10 Arlington Court
11 Valley of the Rocks
12 Lydford
13 Haldon Woods
14 Buckfastleigh
15 Start Point and Berry Head to Sharkham Point NNR
16 Stoke Woods
17 Offwell
18 Exmoor National Park
19 Bracket's Coppice
20 Hooke and Higher Kingcombe
21 The Fleet
22 Portland
23 Wareham Forest and Ringwood Forest
24 Durlston Head and Country Park
25 Arne RSPB Reserve
26 Brownsea Island
27 West Horrington
28 The Quantocks
29 Brean Down
30 Gordano Valley NNR
31 Willsbridge Valley
32 Mells Valley
33 Savernake Forest

1 Isles of Scilly **Grid ref: SV905104 (Hugh Town)**

Key species

Common Dolphin, Lesser White-toothed Shrew, Grey Seal, Common Pipistrelle.

Access

The *Scillonian III* passenger ferry runs from late March through to early November. The timetable varies throughout this period and it would be advisable to seek more information at the Isles of Scilly Travel website: www.islesofscilly-travel.co.uk. The Isles can also be accessed via helicopter out of Penzance (www.islesofscillyhelicopter.com) or by air-bus from the Land's End airfield near St Just (www.islesofscilly-travel.co.uk/air).

Site description

Common Dolphin: The *Scillonian III* has a long history of encounters with Common Dolphins. The waters between Land's End and the Isles of Scilly are a rich fishing area for cetaceans and for the fishermen who flock here from local ports and further afield. Sightings of this species can occur from any of the islands: schools of up to 500 have been recorded from St Mary's. In recent times, specially run pelagic boat trips for birders during the summer out of St Mary's quay have yielded many observations throughout the Western Approaches.

Lesser White-toothed Shrew: One of the best places to look for this species is on the rocky beach at Porthloo, St Mary's. At the top of the beach carefully turn over some of the larger boulders and rocks under which the shrews shelter. Alternatively walk quietly around the top of the beach listening for their shrill high-pitched calls. Once you locate them by call, sit nearby and watch for shrews running between large boulders. The stretch of beach close to Hugh Town, particularly the area backed by tamarisk bushes, is a good place to start searching as it has produced plenty of sightings in the recent past. The sandhoppers that occur on such sandy beaches are a favoured prey item.

The only other land mammals on Scilly are the **Hedgehog**, **House Mouse**, **Wood Mouse**, **Brown Rat** and **Rabbit**. The **Common Pipistrelle** is the only confirmed breeding bat species.

Grey Seal: The Isles of Scilly are a major breeding ground of the Grey Seal. Breeding areas on the archipelago include St Martin's and the Eastern Isles. Specially organised boat trips are available to view this species through the summer.

2 Penwith Peninsula, Cornwall

Grid ref: SW348318 (Cape Cornwall)

Key species

Bottle-nosed Dolphin, Common Dolphin, Risso's Dolphin, Killer Whale, Minke Whale, Harbour Porpoise, Grey Seal.

Access

Britain's south-westernmost mainland point, the Penwith Peninsula, is accessed through Penzance, with roads serving all of the sites given below.

Site description

Bottle-nosed Dolphin: The waters off the Penwith Peninsula, at the far south-western end of Cornwall, are home to a semi-resident school of Bottle-nosed Dolphins that returned to the area in 1991 after a long absence. This group can turn up off any of the headlands on the peninsula between Mount's Bay (off Penzance) and St Ives, but are primarily seen off Cape Cornwall (accessed along minor roads west from St Just) and Land's End. The dolphins are seasonal, spending the winter in southern Cornwall and moving north during spring and summer.

Common Dolphin: This area is one of the best sites in mainland Britain for sightings of Common Dolphins. Although this species is predominately pelagic in nature, with regular sightings from the Isles of Scilly passenger ferry, sightings from Penwith are fairly common. Pendeen Watch and Porthgwarra usually produce the majority of observations.

Risso's Dolphin: Small schools of this species are seen almost annually from headlands on the Penwith Peninsula. Late summer and autumn appear to be the best times, with irregular sightings from Porthgwarra and Land's End.

Killer Whale: Late summer and autumn produces regular Killer Whale sightings off south-west Cornwall. Although infrequent, pods and/or individuals have been seen from Land's End, Lamorna and Cape Cornwall in recent years. The *Scillonian* passenger ferry that runs from Penzance to St Mary's also reports Killer Whales throughout the summer.

Minke Whale: Although the chances of success are slim, Minke Whales can sometimes be spotted from headlands on the peninsula. Late summer and autumn are the prime times as individuals commence their return migration from northerly latitudes past the most westerly points of mainland Britain. Minke Whales have been sighted from promontories such as Land's End, Porthgwarra and Cape Cornwall.

Harbour Porpoise: Sea-watching from any promontories on the peninsula provides opportunities to see Harbour Porpoises. Productive headlands include Cape Cornwall and Land's End.

Grey Seal: Grey Seals can often be observed off any point of the Penwith Peninsula. Around Land's End there are a number of good haul-out sites on rocks just offshore, where the species can be readily observed. Depending upon wind and tide conditions there is often the chance of watching individuals basking for several hours at a time. Cape Cornwall is another favoured haunt.

3 Five Acres CWT Reserve, Cornwall

Grid ref: SW793485

Key species

Water Shrew.

Access

Take the B3284 north from Truro and the right fork, signposted to Perranporth, at the junction in Allet. The Five Acres reserve is immediately on the right. There is ample parking just inside the entrance or alternatively in the triangle between the B3284 and entrance to the nature reserve. All paths, including the pond platform, are suitable for wheelchair access. Access is free.

Site description

The headquarters of the Cornwall Wildlife Trust lie within the grounds of the reserve. This small demonstration reserve includes, among other things, the BBC TV Ground Force wildlife garden and a small population of **Water Shrews** that inhabit the ponds. Other species found on the reserve include **Pygmy** and **Common Shrews**.

4 Cardinham Woods, Cornwall

Grid ref: SX100667

Key species

Hazel Dormouse, Roe Deer, Red Deer.

Access

Cardinham Woods are just east of Bodmin, south of the village of Cardinham. From Bodmin, take the A38 towards Liskeard and then follow the brown tourist signs to Cardinham Woods. There are toilets in the car park. The paths within this 250-hectare mixed woodland are generally firm, although they can become slippery after periods of wet weather.

Roe Deer (*F. Müller*)

Site description

Cardinham Woods cloak a steep valley that runs down to the Cardinham Water River.

Hazel Dormouse: The Hazel Dormice within Cardinham Woods are subject to a long-term monitoring scheme in the Callywith Wood Wildlife Research Area. This species is obviously difficult to spot but it is worth spending time in Callywith Wood, the first section of woodland you come through as you walk north from the parking area.

Other species that may be encountered with a bit of luck include **Red Deer**, **Roe Deer**, **Otters** and **Badgers**.

5 Prideaux Wood, Cornwall **Grid ref: SX064554**

Key species

Greater Horseshoe Bat.

Access

The reserve can be accessed from St Austell through St Blazey on the A390. Just before the railway line take a left turn before turning right after a quarter of a mile. The entrance is shortly after this junction, on the right. There is limited parking on the nearby lanes. Several paths cross the reserve, but they are uneven and may be very muddy after wet weather. Visitors are asked to stay on the paths and be aware that there are many mine shafts in the area.

Site description

This Cornish Wildlife Trust reserve contains one of the few known maternity roosts of **Greater Horseshoe Bats** in Britain. They breed within the derelict mine shafts, which are also used as the winter hibernation site. Parts of the wood consist of ancient broadleaved woodland, although much of it was planted up with coniferous trees during the 1960s. The Wildlife Trust is gradually removing the coniferous trees in order to benefit this important bat population.

6 River Camel, Cornwall **Grid ref: SW987723 (Wadebridge)**

Key species

Otter.

Access

Anywhere where the river is accessible between Wadebridge and Padstow is recommended. The Camel Trail follows the river from Wenfordbridge to Padstow, mainly following a disused railway track, and there is open access all year. The path is suitable for wheelchairs and pushchairs.

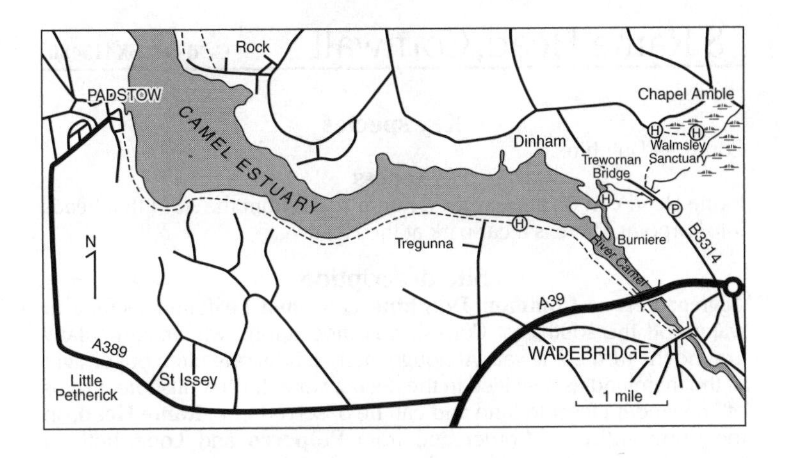

Site description

The River Camel is the main stronghold of the **Otter** in south-west England. Surveys have indicated a dense population along this river, which survived when many others were wiped out during the 1960s and 1970s. It provided a nucleus from which many nearby river systems were recolonised. The wooded lower reaches of the river provide excellent habitat for resting and breeding.

7 Minster Church SSSI, Cornwall

Grid ref: SX110904

Key species

Greater Horseshoe Bat.

Access

Minster Church is situated on the north Cornwall coast and can be reached via the minor road between the villages of Boscastle and Lesnewth.

Site description

The **Greater Horseshoe Bat** colony at Minster contains up to 200 individuals and is the largest known maternity roost in Cornwall and one of the largest in the UK. This colony forages in the Valency Valley, where much of the grassland over which the species feeds is on National Trust property and is maintained by traditional grazing methods. The bats can be seen from the churchyard, or in the surrounding area at dusk.

8 Rame Head, Cornwall **Grid ref: SX415480**

Key species

Common Dolphin.

Access

Rame Head can be accessed via minor roads from the A374 that heads into Torpoint. There is a car park at the Head.

Site description

Aggregations of **Common Dolphins** can often be found feeding in waters off the south-east Cornish coastline during winter, particularly around the turn of the year. Although many schools are often out of sight of the mainland as they feed in the deeper water further offshore, groups often wander closer to land and can be observed from **Rame Head**, on the Cornwall/Devon border, and from **Polperro** and **Looe**, both in Cornwall. Aggregations can number many hundreds of individuals at this time of year.

9 Lundy, Devon **Grid ref: SS145456**

Key species

Sika Deer, Feral Goat, Harbour Porpoise.

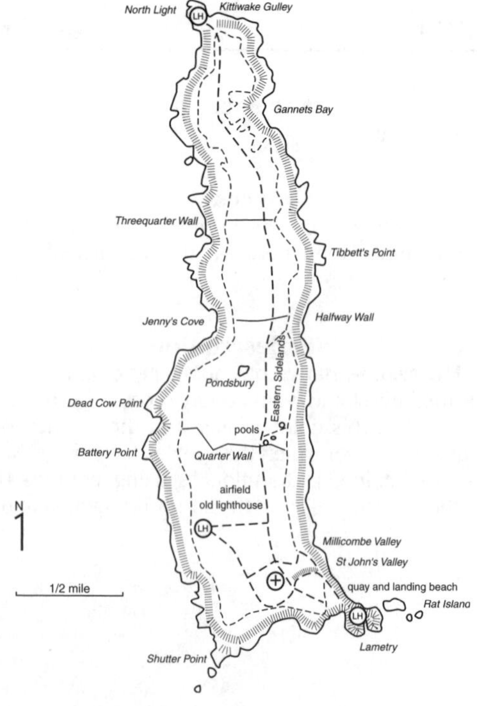

Access

Lundy can be reached by boat and helicopter. For details of accommodation and travel contact:
Lundy Shore Office, The Quay, Bideford, Devon EX39 2LY; tel. 01237 470442; fax 01237 477779; e-mail info@lundyisland.co.uk; website www.landmarktrust.co.uk

Site description

Lundy is a lump of granite, three miles (5.6km) long and half a mile (0.8km) wide, which rises over 120m above the sea where the Bristol Channel meets the Atlantic Ocean.

Sika Deer: A herd of around 40 Sika Deer occur on the island. They tend to be found in the wooded valley areas but are easy to observe with patience, particularly at dusk when their high-pitched whistle is frequently heard.

Feral Goat: During 1991 the herd on Lundy increased in size when another six were brought in from the Valley of the Rocks in Devon. The island is small so viewing the goats should not prove too difficult particularly around the north-east coast of the island.

Black Rat: A colony of Black Rats has recently been extirpated from Lundy due to fears that the drop in breeding seabird numbers had been caused by this nationally rare mammal consuming the eggs of birds such as Manx Shearwaters and Puffins.

The only native terrestrial mammal to be found on Lundy is the **Pygmy Shrew**, which has probably been present since the island separated from the mainland around 8,000 years ago. **Rabbits** and **House Mice** have both been introduced and are now flourishing.

10 Arlington Court, Devon **Grid ref: SS609405**

Key species

Lesser Horseshoe Bat.

Access

Arlington Court is a National Trust property located eight miles (13km) north-east of Barnstaple in the village of Arlington, accessed off the A39. The house is open from 11:00 to 17:00 hrs from the end of March to the end of October, and the grounds stay open every day during winter from dawn until dusk. Entrance is free to National Trust members although otherwise charges of £7 (adults) and £3.50 (children) were in force in 2006. There is free parking and disabled access.

Site description

The 1,125-hectare Arlington Estate lies in the thickly wooded River Yeo valley. From May to September visitors can take the opportunity to view Devon's largest **Lesser Horseshoe Bat** colony via the 'batcam'. For those wanting to see the bats in real life, and not just on a television screen, the woodlands and pastures surrounding the property attract foraging bats during the spring and summer. Deer Park Wood and Woolley Wood are worth exploring via the network of public rights of way.

11 Valley of the Rocks, Devon

Grid ref: SS707498

Key species

Feral Goat.

Access

Situated on the north Devon coast, the Valley of the Rocks runs west from Lynton. There is a car park at the above grid reference and a minor road and public footpath run through the valley from Lynton.

Site description

A herd of up to 40 **Feral Goats** has occurred in this area of Devon for a number of years. In 1991 six of the herd were translocated to Lundy.

12 Lydford, Devon

Grid ref: SX497851

Key species

Hazel Dormouse, Red Deer, Roe Deer, Badger, Otter.

Access

Lydford is on the western edge of Dartmoor just off the A386. From the centre of Lydford village turn right at the war memorial and continue along this road for approximately two thirds of a mile (1km) before turning left onto the forest track.

Site description

Lydford Woods overlook the Lyd Valley and are managed by the Forestry Commission. A circular walk takes you through areas of mixed coniferous trees down to the river and then back to the car park. Due to their nature none of the key species listed are particularly easy to see. However visits early in the morning or late in the evening should provide sightings of **Red Deer**, particularly where extensive views of the area can be obtained over the Lyd Valley and Dartmoor, and **Roe Deer**. **Otters** are infrequent visitors along the river, whilst both **Hazel Dormice** and **Badgers** are typically secretive in areas of favoured habitat.

13 Haldon Woods, Devon

Grid ref: SX884849

Key species

Hazel Dormouse, Brown Long-eared Bat.

Access

Take the Exeter Racecourse turning from the A38 and then follow signs for forest walks. There are car parks at the bird of prey watchpoint and Buller Hill, both of which are signposted.

Site description

Haldon Woods straddle the top of Haldon Ridge, approximately ten miles (16km) south-west of Exeter. Pockets of heathland have been re-established within the forest to complement the vast tracts of broad-leaved and coniferous woodland, much of which is managed by the Forestry Commission.

Hazel Dormouse: The woodlands at Haldon are rich in wildlife including Hazel Dormice, which inhabit the abandoned hazel coppice plots.

Brown Long-eared Bat: There is a chance of observing Brown Long-eared Bats along any of the trails within the woods at twilight, but they are most regularly seen hunting above the Green Walk.

Haldon Woods are home to many other species including a partially melanistic population of **Fallow Deer**, **Roe Deer**, **Stoats**, **Badgers** and both **Common** and **Soprano Pipistrelles**.

Other information

Honey Buzzards are irregularly seen from the watchpoint between mid-May and the end of August.

14 Buckfastleigh, Devon **Grid ref: SX740671**

Key species

Greater Horseshoe Bat.

Access

The village of Buckfastleigh, on the eastern edge of Dartmoor, is easily accessed off the A38 between Exeter and Plymouth.

Site description

A sizeable roost of **Greater Horseshoe Bats** summers in a cave in the grounds of the Abbey Hotel in Buckfastleigh. The added advantage of this site is that up to 400 bats may be watched emerging at night from the comfort of the hotel bar!

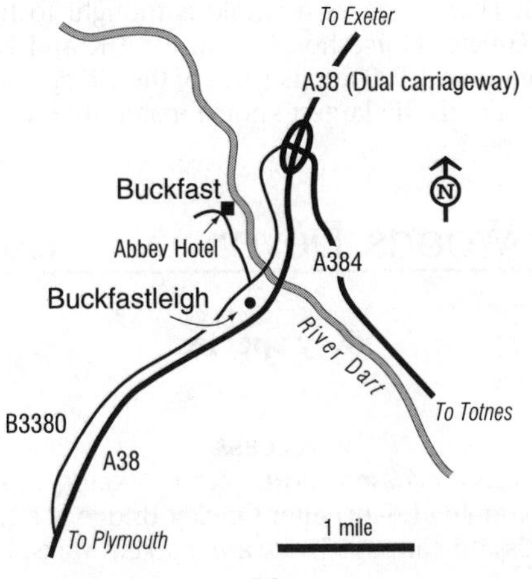

15 Start Point and Berry Head to Sharkham Point NNR, Devon

Grid refs:
Berry Head SX94956
Sharkham Point SX94054
Start Point SX831371

Key species

Harbour Porpoise, Greater Horseshoe Bat, Lesser Horseshoe Bat.

Access

Start Point is one of the most southerly headlands in Devon and is accessed via minor roads from Torcross off the A379. Parking is available by the track beyond Start Farm at SX820374. The Berry Head to Sharkham Point National Nature Reserve is just south of Torbay at Brixham. Both headlands can only be accessed by public footpath from either the car parks listed below, or via the South West Coast Path. There is a country park at Berry Head with parking at SX940562. Parking for Sharkham Point is at the end of St Mary's Road at SX931547.

Site description

Harbour Porpoise: The winter months from November onwards offer the best chance of seeing small groups of porpoises from these headlands. Start Point is usually the most productive of the three, although sightings are still frequent at Berry Head and animals are occasionally seen in Tor Bay.

Greater and Lesser Horseshoe Bats: Berry Head to Sharkham Point National Nature Reserve is a 62-hectare English Nature managed nature reserve that consists of caves, grassland and coastal habitats at the south end of Tor Bay. Berry Head is an extensive outcropping of Devonian limestone that has been drastically altered on its north flank by quarrying. The resulting caves now provide breeding roosts and hibernacula for both **Greater** and **Lesser Horseshoe Bats**. The population of Greater Horseshoe Bats numbered around 70 individuals in the late 1990s but they were declining. Since the introduction of cattle, dung beetles have slowly colonised, in turn benefiting the bats and increasing their numbers. The South Hams area as a whole is thought to hold the largest population of Greater Horseshoe Bats in the UK and is the only area containing more than 1,000 adults (31% of the UK species population). The area also contains the largest known maternity roost in the UK and possibly in Europe.

16 Stoke Woods, Devon

Grid ref: SX919959

Key species

Roe Deer.

Access

Stoke Woods are accessed north from Exeter along the A396. Approximately a mile (1.6km) after Cowley Bridge is a right turn to the Forestry Commission car park. There are marked walks.

Site description

Stoke Woods offer refuge for many mammal species including **Roe Deer**, which are most likely to be seen at dawn and dusk as they feed along the ride and track edges. Other species that occur here include **Brown Long-eared Bats**, **Hazel Dormice** and **Badgers**. The woods are on a steep hillside and a seating area at the top allows fantastic views over the surrounding region.

17 Offwell Woods, Devon **Grid ref: SY185998**

Key species

American Mink.

Access

Offwell is approximately three miles (nearly 5km) east of Honiton and is accessed via minor roads off the A35. To get there turn right at the church in Offwell village. Go down the hill and park along the road. From here, walk back towards the church and turn left to follow the road down the hill. After a few hundred yards and immediately past the last house on the right-hand side of the road, there is a public footpath across a field. Access to the site is through a kissing gate opposite.

Site description

The woods at Offwell are extremely diverse with deciduous and coniferous woodland interspersed with areas of heathland and many small pools. The wood is owned by the Forestry Commission but is managed primarily as a nature reserve in conjunction with the Offwell Woodland and Wildlife Trust. **American Mink** occur frequently at Offwell and may be found hunting the areas of water or, alternatively, in the sections of heathland. Other species here include **Hazel Dormice**, **Red Foxes** and **Roe Deer**. **Otters** are occasionally seen.

18 Exmoor National Park, Devon/Somerset **Grid ref: SS890418**

Key species

Red Deer, Barbastelle, Bechstein's Bat.

Access

Dunkery Beacon is approximately ten miles (16km) south-west of Minehead and is accessed off the A396 via minor roads at Wheddon Cross. There is parking at SS904419.

Site description

The 692km^2 of Exmoor National Park support a diverse array of habitats including heathland, coastline, ancient oak woodland and old beech hedge banks. Over 10,500 people live within the park but the area is still relatively unspoilt, even though it attracts more and more visitors every year.

Hedgehog (*V. Ree*)

Red Deer: Exmoor National Park supports a large population of this species estimated in 1991/92 to be approximately 4,750 individuals. They have survived within the region since prehistoric times. This is unique for this species in England as other populations became extinct through hunting and have since become re-established. Although herds are nomadic, they do favour the higher hills where heather is the dominant vegetation type. There are several minor roads crossing the area and several car parks available for use. The hillsides around Cloutsham and Dunkery Beacon usually produce sightings, although the species may be encountered almost anywhere.

Barbastelle and Bechstein's Bat: Maternity colonies of both Barbastelles and Bechstein's Bats utilise a range of tree roosts in this area of oak woodlands. The National Nature Reserve of Horner Wood (SS893442) may be accessed south from the A39 north Devon/Somerset coast road through the village of Horner. There is a public bridleway as well as various tracks that run through the wood.

Exmoor National Park supports a total of 30 breeding mammals, a total that includes **Roe Deer**, **Fallow Deer**, **Hedgehogs**, **Hazel Dormice** and **Badgers**.

Other information

A strong population of a nationally rare butterfly, the Heath Fritillary, may be found in the area.

19 Bracket's Coppice, Dorset

Grid ref: ST514074

Key species

Bechstein's Bat.

Access

Bracket's Coppice is within a Special Area of Conservation. The woodland is five miles (8km) south-east of Crewkerne near Higher Halstock Leigh. There is access and limited parking along the farm track and permissive footpath opposite Wynford Cottage on the Halstock to Corscombe road. There are numerous good paths throughout the reserve. Please keep dogs under control.

Site description

This 56-hectare site contains one of the first maternity colonies of **Bechstein's Bats** to be discovered in the British Isles. The site consists mainly of broadleaved deciduous woodland with areas of grassland, and the bats utilise specially erected batboxes in the area. **Please remember that bat boxes cannot be checked without a licence**.

20 Hooke and Higher Kingcombe, Dorset

Grid ref: SY533989 (Higher Kingcombe)

Key species

Wild Boar.

Access

The villages of Hooke and Higher Kingcombe are located between Beaminster and Maiden Newton.

Site description

This population, thought still to number less than 100 individuals, is thought to originate from a now defunct **Wild Boar** farm in Bridport. Escapees from another farm operating in the area have recently supplemented numbers. They inhabit the woodland in and around Powerstock Common. There are public footpaths through these woodlands. Please note that sows with piglets can be extremely dangerous so care must be taken.

Other species

Fallow Deer, **Roe Deer** and **Badgers** all occur on Powerstock Common. Great Crested and Smooth Newts both occur in the ponds on the common.

21 The Fleet, Dorset

Grid ref: SY575840

Key species

Water Vole.

Access

The Fleet is a large lagoon on the Dorset coast west of Weymouth. There are various access points as footpaths run down to the water along most of its entire length. A good place to start is the Abbotsbury Swannery, which is well signposted off the B3157 coast road between Bridport and Weymouth. It opens seven days a week from mid-March through to the end of October each year.

Site description

Most famous for the nesting Mute Swans, the swannery at the western end of The Fleet was established as a sanctuary area by monks in 1393. **Water Voles** thrive in the area and may be observed at other locations along this stretch of coast, including the lagoons at West Bexington. The footpaths that lead down to The Fleet from the village of Langton Herring are also good spots for this species.

Other information

There are further car parks and an information centre at Ferry Bridge. The sanctuary area is closed at all times and part of Chesil Beach is closed during the breeding season.

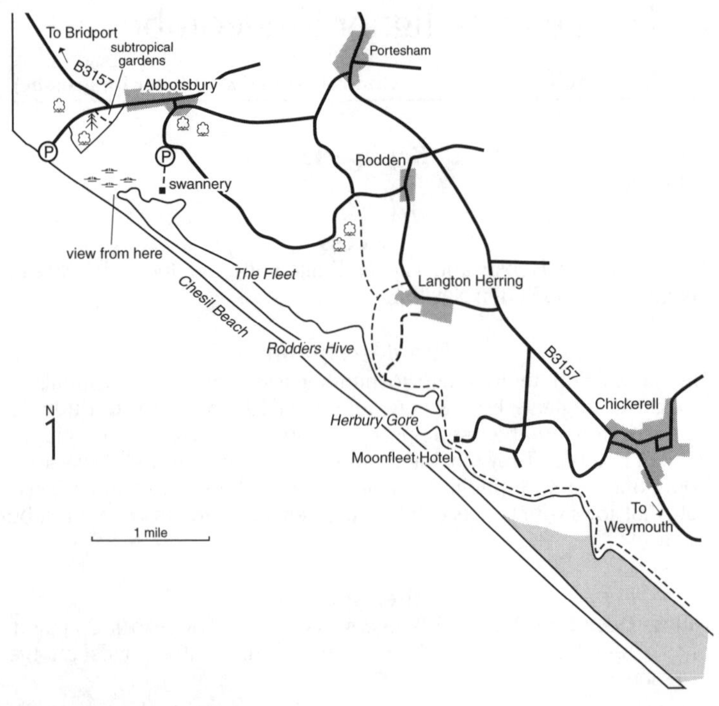

22 Portland, Dorset

Grid ref: SY677685

Key species

Bottle-nosed Dolphin.

Access

Portland Bill, the southern tip of the Isle of Portland, is approached from Weymouth on the A354.

Site description

Small numbers of **Bottle-nosed Dolphins**, possibly from along the coast at Durlston, have started to be seen more frequently off Portland Bill and in West Bay between here and Bridport, off Chesil Beach.

Other information

The latest bird and wildlife information can be obtained from the Portland Bird Observatory website: www.portlandbirdobs.btinternet.co.uk

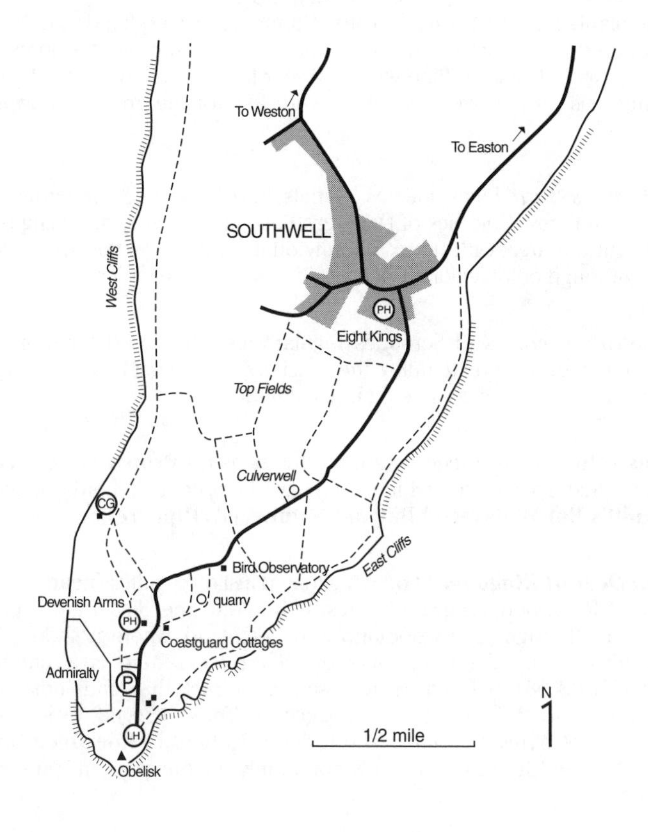

23 Wareham Forest and Ringwood Forest, Dorset

Grid ref: SY891924 (Wareham Forest), SU108058 (Ringwood Forest)

Key species

Serotine, Natterer's Bat, Sika Deer, Roe Deer.

Access

Wareham Forest covers an extensive area north-east of Poole Harbour and can be accessed via minor roads off either the A352 or A35. There is parking at Stroud Bridge (SY889916), opposite Decoy Heath off the B3075 (SY921910), and at SY826903 between Bovington Camp and Clouds Hill. There are numerous trails through the open woodland and heaths that now constitute much of this maturing forest.

Site description

First planted in 1929 this Forestry Commission woodland has been monitored for its bat populations since 1975. At present, the forest is largely (60%) Corsican Pine with a 60-year felling cycle. A further 15% is continuous broadleaved woodland with 25% of the area being open heath.

Natterer's Bat: Eight species of bats have now been recorded in Wareham Forest. Colonies of Natterer's Bats have been found using bat-boxes in the area, with up to 55 individuals found together in a single box, although colonies of 30–40 individuals are more typical.

Serotine: Colonies of Serotines regularly use this forest for foraging although they have yet to utilise the specially erected bat boxes that have attracted other species since their instalment.

Pipistrelle species breed within the forest, as do **Brown Long-eared Bats**. Other species for which there are occasional records include **Brandt's Bat**, **Whiskered Bat** and **Nathusius's Pipistrelle**.

Sika Deer at Ringwood Forest: Situated just three miles (nearly 5km) west of Ringwood, Ringwood Forest is home to the 'Sika Trail', a path that runs through pine woodland and heathland, where it should be possible to spot good numbers of Sika Deer. There is a parking area (SU108058) that can be accessed via a track that leads north off the minor road between the villages of Three Legged Cross and Ashley Heath. This trail may also produce sightings of **Roe Deer**. Both Sika Deer and Roe Deer are also commonly encountered in Wareham Forest.

24 Durlston Head and Country Park, Dorset

Grid ref: SZ037772

Key species

Bottle-nosed Dolphin.

Access

Durlston Head is just south of Swanage on the Isle of Purbeck in south Dorset. The area is accessed via a minor road south from Swanage and there is parking at Durlston Country Park. Other facilities here include toilets, a visitors' centre and a café.

Site description

Bottle-nosed Dolphin: Durlston Country Park was established by Dorset County Council in the 1970s and incorporates a diverse range of habitats within its 100 hectares. Durlston Head has been home to the 'Dolphin Watch' since 1988, and a semi-resident population of up to six animals has been identified along the coast. There are two peak times to see this species off the Purbeck coast: April–May and September–October, although the spring peak has been known to shift to March in some years. During these periods, the dolphins can be seen on approximately 80% of all days, at distances of 30–300m from the shore.

Other species that occur within the country park include **Roe Deer**, which are commonly seen early in the morning, with the gully a particularly favoured location. There are also **Red Foxes**, **Common Shrews**, **Field Voles**, **Stoats**, **Weasels**, **Badgers**, **Common** and **Soprano Pipistrelles** and **Noctules**. Both **Common** and **Grey Seals** are irregular visitors to the coastline.

Other information

The nationally rare Early Spider Orchid may be found within the short-turf areas within the country park. Other orchids include Autumn Lady's Tresses, Green-winged Orchid and Pyramidal Orchid. A total of 33 butterfly species breed within the country park: notable species include Lulworth Skipper, Adonis Blue and Chalkhill Blue.

25 Arne RSPB Reserve, Dorset

Grid ref: SY969876

Key species

Sika Deer, Roe Deer, other common mammals.

Access

Arne RSPB Reserve is east of Wareham off the A351 Swanage road. Either take the turning signposted to Arne or the road from Stoborough village. The Shipstal and Coombe Birdwatchers' Trails are open to visitors at all times. There is a small car parking charge for non-members and the gate is locked at dusk.

Site description

The reserve is home to a large population of Sika Deer: indeed, the Isle of Purbeck (upon which Arne is sited) is now home to the largest group of Sika Deer in England. They are extremely easy to see at Arne as the reserve encompasses large areas of open habitat such as lowland heathland and saltmarsh. Both are prime foraging areas for the deer. They are particularly conspicuous during the rutting season in October. **Roe Deer** are less frequently encountered but may be seen early and late in the day along the woodland edges. Other mammals in the reserve include **Harvest Mice**, **Badgers** and **Red Foxes**.

Other information

This is one of the few places in Britain where all six native reptiles occur.

26 Brownsea Island, Dorset **Grid ref: SZ016879**

Key species

Red Squirrel, Sika Deer, Water Vole.

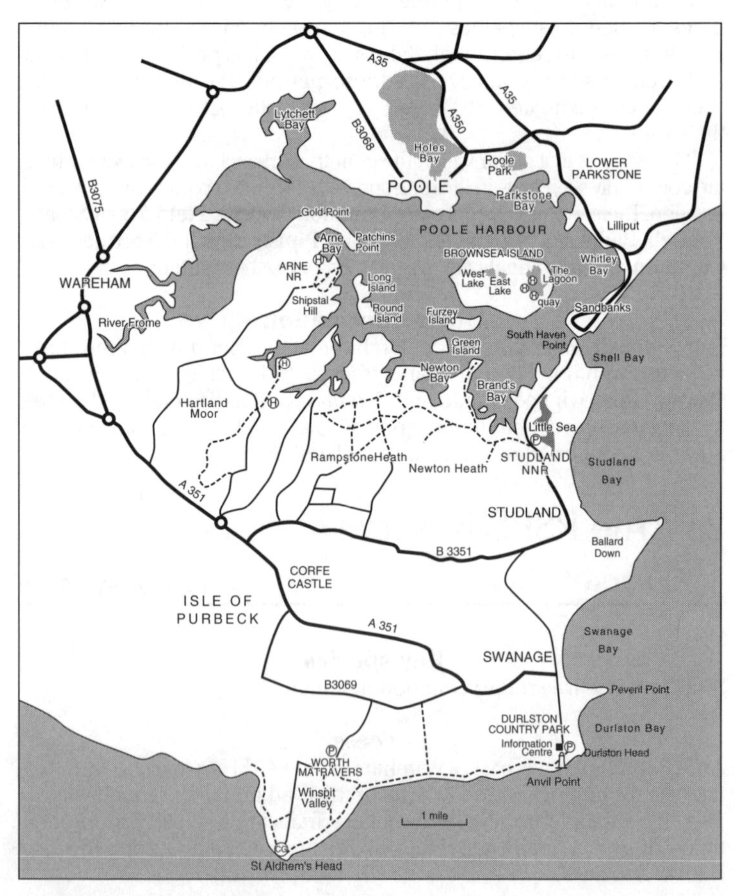

Access

Brownsea Island is a National Trust property located within Poole Harbour in south Dorset. Ferries leave from Poole Quay and Sandbanks, and the island is open from March/April until October/November. There is a small charge to enter the reserve and the National Trust charges a small landing fee on top of the ferry fare.

Site description

Situated less than half a mile from the mainland village of Sandbanks, the woodlands, heathland and lagoons of Brownsea Island offer refuge for several mammal species.

Sika Deer: Sika Deer were introduced to Brownsea from Japan in 1896 but the population was wiped out during the terrible fire of 1934 when virtually all the deer either swam to the mainland or perished. During the 1970s some swam back and numbers have since built back up. Unfortunately they do considerable damage to young trees on the island.

Red Squirrel: Brownsea Island is the largest and the most accessible of the three island refuges within Poole Harbour that provide sanctuary for the local populations of Red Squirrels. They may be found in all the wooded parts of the island and are most active during spring and autumn, at dawn and at dusk. A good time to watch them is in early spring before the leaves of the deciduous trees have developed. Autumn can also provide productive viewing as the squirrels forage on the woodland floor for fallen seeds and fruit. There are approximately 200–250 Red Squirrels on the island.

Water Vole: There are two freshwater lakes at the north end of the wet central valley. With luck and a bit of patience, visitors may observe Water Voles swimming across the channel in front of the reedbed hide.

Wood Mice, **Roe Deer** and **Fallow Deer** also occur on the island.

27 West Horrington, Somerset **Grid ref: ST572474**

Key species

Serotine.

Access

The small village of West Horrington is situated near Wells. Take the B3139 from Wells towards Midsomer Norton. At the first crossroads take the minor road north towards West Horrington. As you enter the village look for a public telephone box on your right-hand side. Park carefully, remembering to respect the privacy of the residents at all times.

Site description

Up to 50 **Serotines** emerge from the gable ends of the property directly behind the telephone box, offering a tremendous experience on a summer evening.

28 The Quantocks, Somerset **Grid ref: ST115425**

Key species

Red Deer.

Access

Although **Red Deer** can be seen anywhere within the Quantocks a good place to start is the Forestry Commission woodland at St Audries which can be accessed off the A39, 2.5 miles (4km) to the east of Williton. The entrance to the wood is signposted at the main gateway.

Site description

A survey of the area in 1989/90 revealed a population of 800–900 **Red Deer** in this range of hills in northern Somerset. A particularly good place to observe them is at St Audries on the northern edge of the hills. The area holds a large population of Red Deer and the rest of the ridge is easily accessed from here. **Roe Deer** also occur.

29 Brean Down, Somerset **Grid ref: ST287591**

Key species

Feral Goat.

Access

This National Trust property juts out into the Bristol Channel just south-west of Weston-super-Mare. It is open all year round and there is parking at Brean Down café at the bottom of Brean Down. National Trust members must display their membership cards. The higher down is a steep climb from the car park.

Site description

A herd of around 15–20 goats is easily seen as the goats patrol the cliff tops and grassland. The area also supports **Badgers**, **Red Foxes** and **Weasels**.

30 Gordano Valley NNR, Avon **Grid ref: ST439735**

Key species

Greater Horseshoe Bat, Serotine.

Access

Gordano Valley is near Walton-in-Gordano and Weston-in-Gordano. The best access to the site is along the B3124, and then onto Walton Drove. Access is limited to public rights of way.

Site description

The Gordano Valley is home to a number of bat species, including **Serotine** and **Greater Horseshoe Bats**, throughout the summer, and both species can be observed from the network of public footpaths that run through the valley.

31 Willsbridge Valley, Avon **Grid ref: ST666707**

Key species

Greater Horseshoe Bat, Noctule.

Access

The reserve lies on the south-eastern edge of Bristol and can be reached from the A431 Bristol to Bath road, turning into Long Beach Road. There is a car park on the left. There are also buses from Bristol to Long Beach Road or Willsbridge Hill. To access the reserve take the public footpath from Long Beach Road.

Site description

This is an Avon Wildlife Trust Local Nature Reserve (LNR) that lies on the edge of Bristol and is surrounded by housing estates. It is home to Willsbridge Mill, the Wildlife Trust's Education and Countryside Centre. During summer, small numbers of both **Greater Horseshoe Bats** and **Noctules** may be observed hunting insects in the valley. Other mammals present include **Badgers** and **Red Foxes**.

32 Mells Valley, Somerset **Grid ref: ST657476**

Key species

Greater Horseshoe Bat.

Access

The Mells Valley lies approximately five miles (8km) north-east of Shepton Mallet, centred on Mells village. Minor roads criss-cross the area so the lack of footpaths should not deter keen bat hunters.

Site description

The Mells Valley SAC (Special Area of Conservation) contains an exceptional breeding population of **Greater Horseshoe Bats**. The area, which is mainly grassland with some deciduous woodland, includes a maternity site containing 12% of the UK Greater Horseshoe Bat population. A proportion of the population also hibernates at the site. Other bat species that have been recorded here include **Lesser Horseshoe Bats**, **Natterer's Bats**, **Noctules**, **Daubenton's Bats** and **Common Pipistrelles**.

33 Savernake Forest, Wiltshire **Grid ref: SU231668**

Key species

Daubenton's Bat, Natterer's Bat, Brandt's Bat, Whiskered Bat, Brown Long-eared Bat, Fallow Deer, Yellow-necked Mouse.

Access

The forest is just south-east of Marlborough and is accessed from either the A396 or the A4. Although private throughout, the Trustees grant extensive public access even though there are no vehicular or public rights of way anywhere within the forest. Except on one day when all tracks are

Yellow-necked Mouse (*P. Barrett*)

barred, to maintain its private status and to prevent the creation of permanent vehicular rights of way, vehicles are permitted to drive down the main avenues within the forest and pedestrians may walk where they like. Camping is only permitted at the Forestry Commission campsite in the north-west corner of the forest, about half a mile south of Marlborough.

Site description

Savernake Forest is the only privately owned forest in Britain. It currently covers 1,800 hectares with many of the timber rights being leased to the Forestry Commission. Most of the area is made up of broadleaved trees, particularly Beech.

This ancient forest is full of wildlife and is particularly attractive to bat enthusiasts who come to watch the swarming activity outside the main entrance to Savernake Tunnel, located on a disused railway, during spring and autumn. An incredible total of thirteen bat species have been recorded in just a mile of forest.

The tunnel on the old railway line connecting Andover and Swindon has vandal-proof walls at each end and grilles have been fitted to allow access for the bats. Savernake Tunnel is principally a winter hibernaculum, although bats can be observed swarming around the north entrance in spring and autumn. The tunnel houses a nationally important wintering colony of **Natterer's Bats**. During the 29 winter surveys completed up to April 2003, 1,997 bats were recorded. Of these 93% were Natterer's Bats. This makes the tunnel one of the most important sites in the world for this species. **Daubenton's Bat** is the second most commonly recorded species since recording began in January 1993. It is also the third most common species during swarming activity in the autumn. **Brown Long-eared Bats** are commonly encountered and clusters are often found during bat-box checks. Serotines are regularly seen over the forest and **Barbastelles**, **Brandt's Bats** and **Whiskered Bats** are also occasionally encountered while odd **Bechstein's Bats** have started to appear recently.

Fallow Deer are common within the forest, while **Roe Deer**, **Muntjac** and small numbers of **Red Deer** are also resident. The rodent fauna includes **Hazel Dormice**, **Wood Mice** and **Yellow-necked Mice**. **Red Foxes**, **Badgers** and common mustelids such as **Stoats** and **Weasels** are also present.

Other information

The Mammals Trust UK organises a trip each year to help the Wiltshire Wildlife Trust inspect the nest boxes and carry out mist-netting of bats at the tunnel entrance (numbers are strictly limited). For details please refer to the Mammals Trust UK website: www.mtuk.org.uk.

SOUTH-EAST ENGLAND

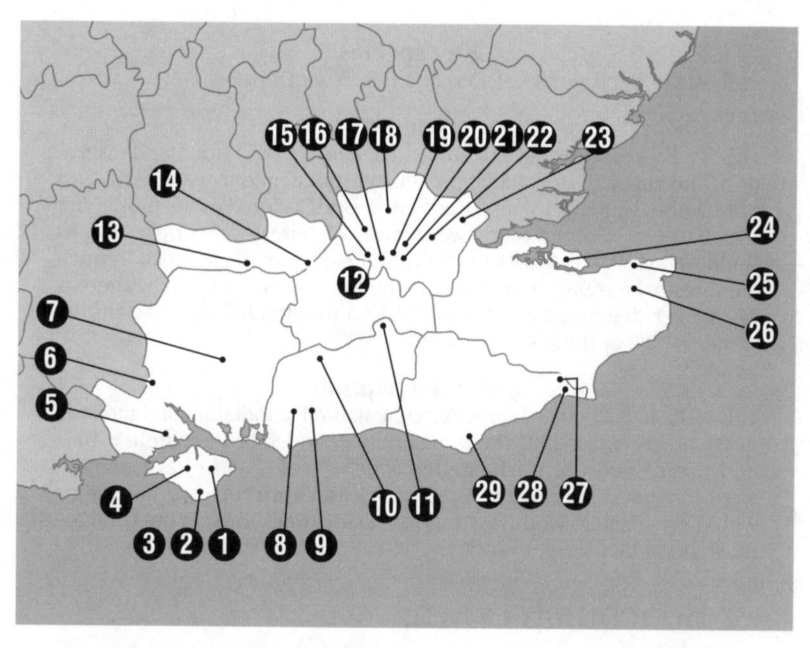

1 Briddlesford Woods
2 Bonchurch Down
3 Isle of Wight woodlands
4 Carisbrooke Mill Pond
5 New Forest
6 Mottisfont Abbey
7 Micheldever Wood
8 Singleton and Cocking Tunnels
9 The Mens
10 Ebernoe Common
11 Chiddingfold Forest
12 Mole Gap to Reigate Escarpment
13 Bowdon Woods
14 Windsor Great Park
15 Bedfont Lakes Country Park
16 Grand Union Canal
17 Bushy Park
18 Highgate Wood
19 Wimbledon Common
20 London Wetland Centre
21 Beddington Park
22 Wanstead Flats
23 Berwick Ponds
24 Elmley Marshes RSPB Reserve
25 Herne Bay
26 Stodmarsh NNR
27 Kent/East Sussex border
28 Rye Harbour Nature Reserve
29 Abbot's Wood

1 Briddlesford Woods, Isle of Wight

Grid ref: SZ549904

Key species

Barbastelle, Bechstein's Bat, Red Squirrel, Hazel Dormouse.

Access

This is a People's Trust for Endangered Species (PTES) woodland reserve of 167 hectares. It is located approximately three miles (nearly 5km) east of Newport. There is no public access to Briddlesford Woods proper but the Forestry Commission woodland of Firestone Copse flanks Briddlesford Woods on its north-eastern side, providing walks through the forest where bats from the neighbouring woodland forage. There is also a track that runs south from the A3054 Wootton Bridge and skirts the western edge of Briddlesford Woods.

Site description

This is the largest block of ancient semi-natural woodland left on the Isle of Wight. In August 2004 three maternity colonies of **Bechstein's Bats** and two colonies of **Barbastelles** (one of males and one maternity) were identified within the woodland. **Red Squirrels** are flourishing within Briddlesford Woods, as are **Hazel Dormice**: however the latter are obviously a lot trickier to spot!

2 Bonchurch Down, Isle of Wight

Grid ref: SZ572786

Key species

Feral Goat.

Access

This site is just north of Ventnor on the south coast of the Isle of Wight. Access to the parking area is from the minor road that runs west from the B3327 on the north side of Ventnor.

Site description

Bonchurch Down has been home to a small herd of **Feral Goats** since 1992, when a group of nine goats were relocated here, from the Valley of the Rocks in north Devon, for scrub control. The population now numbers around 50 animals.

3 Isle of Wight woodlands

Key species

Red Squirrel.

Access

Vehicle ferries to the island run from Southampton, Portsmouth, Gosport and Lymington.

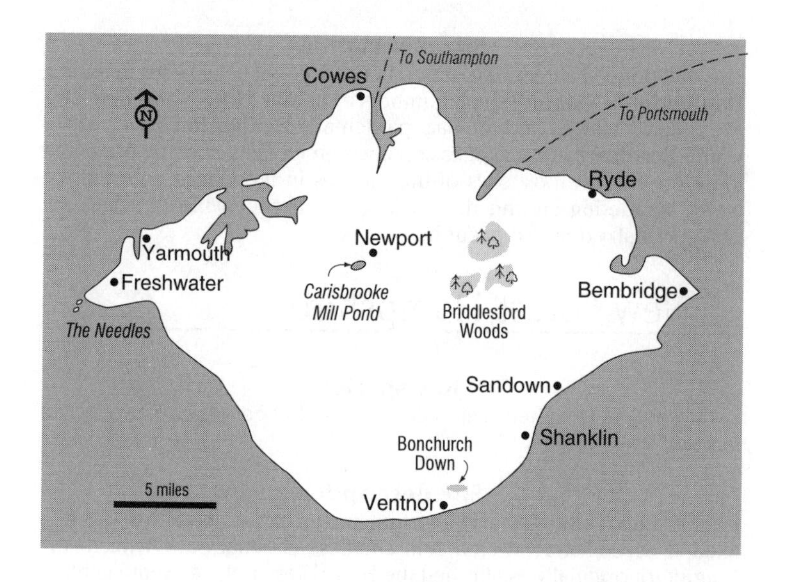

Site description

The Isle of Wight still harbours a large population of around 2,000 Red Squirrels. However, in 2001 small numbers of Grey Squirrels were sighted upon the island. These may have been either accidental or deliberate releases or perhaps they were stowaways on the ferries from the mainland. In 2002 'Operation Squirrel', a partnership of several Isle of Wight organisations was established in order to protect the Red Squirrel population. This involved the following-up of any reports of Grey Squirrels and their subsequent removal if the reports were confirmed. The results of the operation were very successful and no further evidence of Grey Squirrels has been found since. Despite this the risk of future releases is very real and the operation is ongoing.

Red Squirrels are relatively easy to see on the island with strongholds at **Newtown Nature Reserve** (SZ430905), **Walter's Copse** (SZ432905) and **Parkhurst Forest** (SZ472909). They can also be seen at Rylstone Gardens, an area of traditional cliff-top gardens near Shanklin, and at **Alverstone Mead Nature Reserve**, where the squirrels can usually be observed from a hide that overlooks marshes and a pond.

4 Carisbrooke Mill Pond, Isle of Wight

Grid ref: SZ488881

Key species

Lesser Horseshoe Bat, Daubenton's Bat, Serotine.

Access

Carisbrooke is a small village on the western side of Newport town. The pond can be accessed from along Spring Lane.

Site description

The mill pond is attractive to a number of feeding bat species including **Daubenton's Bats** and small numbers of **Lesser Horseshoe Bats**. This site is easily viewed and the bats, particularly **Daubenton's**, are easy to watch. **Serotines** are also occasionally seen foraging over the pond and there are several roost sites of this species in the village, so it can be worth wandering around the back streets between the pub and the castle. **Pipistrelles** also occur here.

5 New Forest, Hampshire

Key species

Red Deer, Sika Deer, Bechstein's Bat, Brandt's Bat, Serotine, Yellow-necked Mouse.

Site description

In 2004 the 578km^2 of rural Hampshire known as the New Forest was designated as England's twelfth, and smallest, national park. William the Conqueror originally established the New Forest in the eleventh century as a hunting area, primarily for deer. Of course, it was originally a very wild place, as vast tracts of broadleaved woodland would have been interspersed with the boggy areas that still prevail today. Since the time of William I the forest has been constantly modified. Indeed, the forest that exists today is completely different from that present late in the eleventh century. One of the major changes resulted from the need for wood during the world wars. In the last century, large blocks have been reforested with fast-growing non-native coniferous tree species.

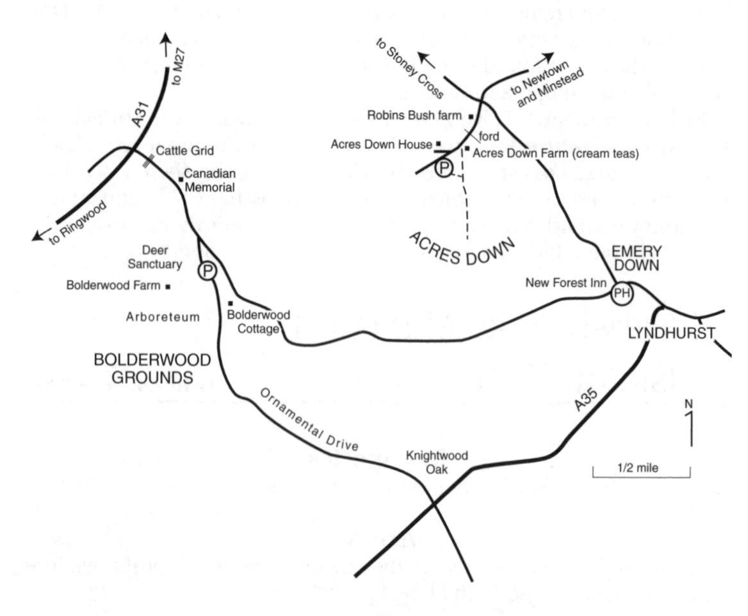

Brandt's Bat (*P. Twisk*)

Many species (particularly butterflies) have declined or become extinct within the forest in the last 200 years, but the New Forest is still a Mecca for naturalists. For mammal watchers, this area has never been better. Surveys in the last two years have located 11 species of bat within the forest, including such rarities as **Bechstein's Bats**, **Grey Long-eared Bats**, **Barbastelles**, **Brandt's Bats** and **Serotines**. Bechstein's Bats breed within Hollands Wood (SU303047), off the A337 between Lyndhurst and Brockenhurst. There is a car park to the west of the main road just north of here at Whitley Wood. Deer continue to be successfully managed with native **Red Deer** and **Roe Deer** now competing for space with the introduced **Sika Deer**, **Fallow Deer** and **Muntjac.** The Red Deer are best looked for around Burley and Brockenhurst, and the Fallow Deer are easiest to find in Bolderwood deer enclosure, where they are fed each afternoon from Easter to September. **Badgers** occur within the forest, although at a relatively low density. Many other woodland mammals, including **Yellow-necked Mice**, may also be found.

6 Mottisfont Abbey, Hampshire

Grid ref: SU325269

Key species

Barbastelle.

Access

Mottisfont Abbey is situated five miles (8km) north of Romsey. Information regarding opening times can obtained from the National Trust website: www.nationaltrust.org.uk, although the gardens and abbey usually close before dusk during the summer.

Site description

Mottisfont Abbey is a National Trust property that has recently been designated as an SSSI and candidate SAC for its **Barbastelle** colony. The abbey's restrictive closing times should not put the keen bat hunter off as much of the surrounding area can be viewed from the minor road that runs west towards Mottisfont village from the A3057. Parking is also available at SU314276 and at SU334270, from where it is possible to explore the surrounding woodlands and countryside for hunting Barbastelles.

7 Micheldever Wood, Hampshire

Grid ref: SU530363

Key species

Brown Hare, Roe Deer, Fallow Deer.

Access

From Junction 9 of the M3, travel north on the A33, before turning right onto Northington Lane. A car park with a picnic area is half a mile down this lane on the left-hand side.

Site description

This predominantly beech woodland, a Forestry Commission site, is home to a number of mammals including **Brown Hares**. They are most likely to be seen among the younger plantations. Sightings of **Roe Deer** and **Fallow Deer** are also frequent, particularly in the Itchen Wood section south of the road.

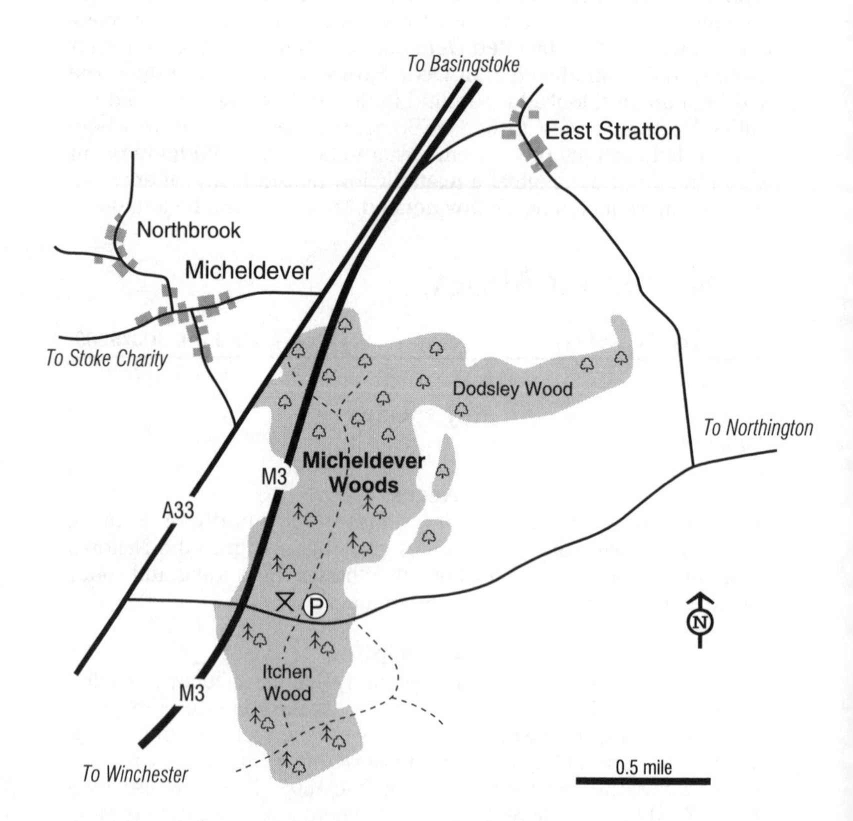

8 Singleton and Cocking Tunnels, West Sussex

Grid ref: SU872144 (central point of a Special Area of Conservation)

Key species

Bechstein's Bat, Barbastelle.

Access

Westdean Woods and Singleton Forests are both large tracts of broadleaved woodland located between the villages of Singleton and Cocking, approximately eight miles (13km) north of Chichester on the A286. Parking is available just south of Cocking at SU873164 and a public bridleway runs west from the A286 just north of Cocking Tunnel into a section of Westdean Woods known as The Marlows. Singleton Tunnel (as well as other abandoned railway tunnels) is just to the north-west of Singleton village and can be accessed via a public footpath north-west from the village pub.

Site description

These tunnel sites on a disused railway line are used as hibernation sites by small numbers of **Barbastelles** and **Bechstein's Bats**. They may be seen, along with other bat species, in swarming activity in the spring and autumn, prior to and after hibernation. Although no summer colonies have yet been found in the area, summer visits may be worthwhile as there are suitable deciduous forests in the vicinity that may hold these species.

Other species recorded in the area and hibernating within the tunnel system complex include **Serotines**, **Brandt's Bats**, **Daubenton's Bats**, **Whiskered Bats**, **Natterer's Bats**, **Brown Long-eared Bats** and **Common Pipistrelles**. This site was also host to a hibernating immature male **Greater Mouse-eared Bat** during the winters of 2002/2003 and 2003/2004. However, subsequent searches during the summer of 2004 failed to find the species.

The woodlands are also home to a variety of other common mammal species including **Water Voles**, **Red Foxes**, **Muntjac**, **Roe Deer**, **Red Deer**, **Fallow Deer**, **Brown Hares** and **Badgers**.

9 The Mens, West Sussex

Grid ref: TQ023236

Key species

Barbastelle.

Access

This Sussex Wildlife Trust reserve lies just off the A272 road between Billinghurst and Petworth. There is a car park at the nature reserve on the road to Hawkhurst Court, and numerous trails, flat but muddy when wet, cut through the woodland.

Site description

The Mens Special Area of Conservation is an extensive area of mature beech woodland that holds small numbers of **Barbastelles**.

10 Ebernoe Common, West Sussex

Grid ref: SU975278 (car park off Streel's Lane)

Key species

Barbastelle, Bechstein's Bat, Grey Long-eared Bat, Whiskered Bat, Daubenton's Bat.

Access

This Sussex Wildlife Trust reserve is approximately five miles (8km) north of Petworth, just south of Ebernoe village. The common is easily accessed off the A823, the best starting point being from the car park next to the church off Streel's Lane. There is an extensive network of footpaths and bridleways within the woodlands and across the common.

Site description

About half of this site is ancient woodland that has been designated as a National Nature Reserve and a Site of Special Scientific Interest, and is a candidate Special Area of Conservation. The woodlands are made up of an extensive block of Beech combined with former wood-pasture. A maternity colony of **Barbastelles** utilises a range of tree roosts in this latter area of old sessile oak woods which contains a dense understorey of holly as well as open glades and open water. Maternity roost sites are usually in dead tree stumps, but the species appears to be present throughout the year, with individuals using a range of roost sites in tree holes and under bark. Barbastelles follow river valleys when foraging and fly down to the wooded chalk hills of the South Downs. They may travel 12 to 19 miles (20–30km) each night.

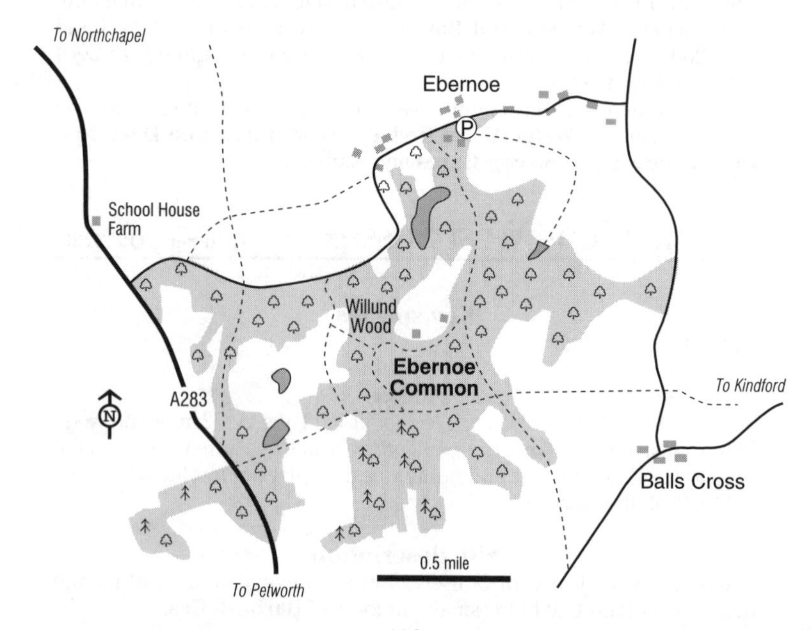

A maternity colony of **Bechstein's Bats** also occurs in the old sessile oak woods. Roosts are mainly in old woodpecker holes in the stems of live mature oak trees. They usually stay around the quarry during foraging missions, rarely straying further than 1–2km from the site. The extremely rare **Grey Long-eared Bat** is also irregularly recorded at this site.

The Mammals Trust UK runs excellent visits to see **Bechstein's Bats** in July. The trips also provide opportunities to see **Barbastelles**, **Whiskered Bats**, **Daubenton's Bats**, **Common Pipistrelles** and **Soprano Pipistrelles**. For details please refer to the Mammals Trust UK website: www.mtuk.org

11 Chiddingfold Forest, Surrey **Grid ref: TQ025351**

Key species

Bechstein's Bat.

Access

Head south towards Horsham on the A281 and turn right at Alford Crossways towards Dunsfold. There is a car park en route to Dunsfold on the left-hand side of the road from where you can enter Sidney Wood and the Chiddingfold Forest.

Site description

Chiddingfold Forest contains the largest area of oak woodlands on the Weald Clay. Much of its 500 hectares have recently been surveyed by the University of Sussex in collaboration with the Natural History Museum, Forest Enterprise and the Mammals Trust UK. At the time of writing three **Bechstein's Bat** maternity colonies with at least 80 breeding females have been identified within the SSSI.

Other information

Butterflies found within the forest include the nationally rare (and declining) Pearl-bordered and Small Pearl-bordered Fritillaries.

12 Mole Gap to Reigate Escarpment, Surrey **Grid ref: TQ199533**

Key species

Bechstein's Bat.

Access

The woodlands around Mole Gap and the Reigate Escarpment are between the towns of Dorking and Reigate, and just south-west of the M25. There are car parks for the woodland at Headley Heath off the B2033 just north of Pebble Coombe.

Site description

This escarpment formed by the deep-cutting of the River Mole is a 888-hectare Special Area of Conservation (SAC) on the North Downs. It supports a wide range of dry calcareous grassland types as well as large areas of broadleaved woodland and smaller tracts of heath and scrub. Small numbers of **Bechstein's Bats** may be found hunting throughout the area in summer. Mole Gap is especially important since it contains the only area of stable Box scrub in the UK. The natural erosion processes provide ideal conditions for this vegetation type.

Other information

The area also supports a wide range of orchids including Musk Orchid and Man Orchid.

13 Bowdown Woods, Berkshire

Grid ref: SU501656

Key species

Roe Deer, Muntjac, Hazel Dormouse, Brown Long-eared Bat, Common Pipistrelle, Soprano Pipistrelle.

Access

The reserve is 2.5 miles (4km) south-east of Newbury and can be accessed from Bury's Bank Road along a track signposted to Newbury Trout Lakes. There is a small car park down this track and there are two further car parks that serve this 54-hectare site. The reserve is open all year round.

Site description

This is a nature reserve managed by the Berkshire, Buckinghamshire and Oxfordshire Wildlife Trust. Small numbers of **Roe Deer** and **Muntjac** may be found within the woodland. **Brown Long-eared Bats** and **pipistrelles** may be watched hunting insects at dusk along the woodland rides whilst **Hazel Dormice** are best located in autumn when they search for food amongst the fruiting shrubs and scrubby plant species.

14 Windsor Great Park, Berkshire/Surrey

Grid ref: SU961724

Key species

Serotine.

Access

The park and forest are easily accessed and well signposted from major roads between Bracknell, Windsor and Egham.

Site description

This vast former Norman hunting chase, set in 6,000 hectares of the Surrey and Berkshire countryside, stretches from Windsor Castle in the north to Ascot in the south. This site has the largest collection of mature, overmature and ancient oak and beech trees remaining in Europe, north of the Mediterranean and the Pyrenees. Such well-established habitats provide ample roosting and foraging areas for a number of bat species. Recent bat surveys have indicated the presence of regular Serotines hunting throughout areas of suitable habitat.

15 Bedfont Lakes Country Park, London

Grid ref: TQ076725

Key species

Nathusius's Pipistrelle, Water Vole.

Access

This Local Nature Reserve is between East Bedfont and Ashford, approximately three miles (nearly 5km) east of Staines. It has good visitor facilities, including disabled paths and toilets.

Site description

The site on which Bedfont Lakes stands once formed a large orchard that supplied Covent Garden up until the 1920s. The area was then worked for sand and gravel in the 1950s and then as a waste disposal tip during the 1970s. After huge renovation and landscaping work by Hounslow Council, the country park was opened to the public in 1995 and gained Local Nature Reserve (LNR) status in 2000. In October 2002 a small number of **Nathusius's Pipistrelles** were found using bat boxes within the park. It remains to be seen whether this will constitute another breeding site for this species, but the presence of a male and female together in one box is certainly exciting.

In the summer of 2002, Mammals Trust UK introduced 200 **Water Voles** with the aim of establishing a further viable population of this species within Greater London.

16 Grand Union Canal, London

Grid ref: TQ053858

Key species

Noctule, Daubenton's Bat, Common Pipistrelle, Soprano Pipistrelle.

Access

Bats occur widely along the canal in the Wembley and Ealing areas. A good place to start is the section just north of the A40 (Western Avenue) along the towpath that runs along the western side of the canal (above grid reference).

Site description

On warm nights during the summer, the section of canal in the London Borough of Hillingdon provides a fine setting to watch **Daubenton's Bats** foraging over the surface of the water. **Noctules** and **pipistrelles** may also be found here, and their differing feeding methods should allow firm identifications to be made.

17 Bushy Park, London **Grid ref: TQ160694**

Key species

Brown Long-eared Bat, Daubenton's Bat, Red Deer, Fallow Deer.

Access

Situated in south-west London between Kingston-upon-Thames and Walton-on-Thames, Bushy Park is accessed south via the footpath from Sandy Lane at TQ159704 or via Hampton Court Gate off Hampton Court Road. There is both disabled parking and toilets available. The park is open from 05:00–22:30 hrs for pedestrians and from 06:30 hrs to dusk for vehicles (19:00 hrs in winter).

Site description

Bushy Park is the second largest of the Royal Parks, covering an area of 450 hectares. **Brown Long-eared Bats** can be observed hunting along the tree-lined avenues at this site during the summer, whilst **Daubenton's Bats** may be watched hunting over the ponds. Other nearby sites, including Hampstead Heath and Beech Hill Lake in Barnet, also offer opportunities to see Daubenton's Bats. Herds of feral **Red** and **Fallow Deer** are usually conspicuous in Bushy Park.

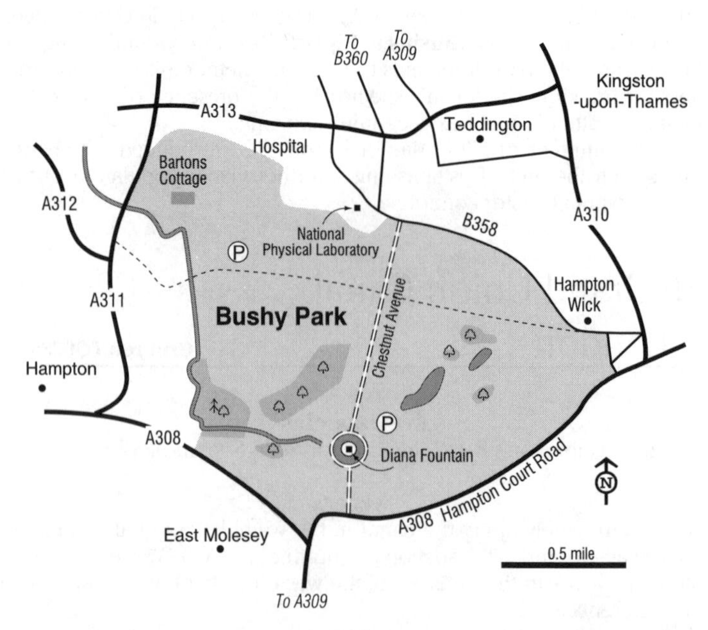

18 Highgate Wood, London **Grid ref: TQ282886**

Key species

Natterer's Bat, Leisler's Bat.

Access

Highgate Wood is in north London approximately three miles (nearly 5km) west of Wood Green at Haringey. The wood enjoys open access throughout. There is a small car park for the use of disabled drivers and there are disabled toilet facilities on site. The paths are suitable for both wheelchairs and pushchairs.

Site description

The woods are made up predominately of oak and other trees, although there are also small numbers of Holly and Wild Service trees present. Small areas of the wood are fenced off to encourage natural regeneration. Seven species of bat, including both **Natterer's** and **Leisler's Bats**, have been recorded at Highgate and several species have roosts in many of the old oaks. Other species present include **Red Foxes**.

19 Wimbledon Common, London **Grid ref: TQ224717**

Key species

Noctule, Serotine, Daubenton's Bat, Brown Long-eared Bat, pipistrelles.

Access

Wimbledon Common is in south-west London. It is divided by the A3 and bounded on the east side by the A219, Wimbledon Parkside. Parking is available at the Windmill Museum with access off the A219.

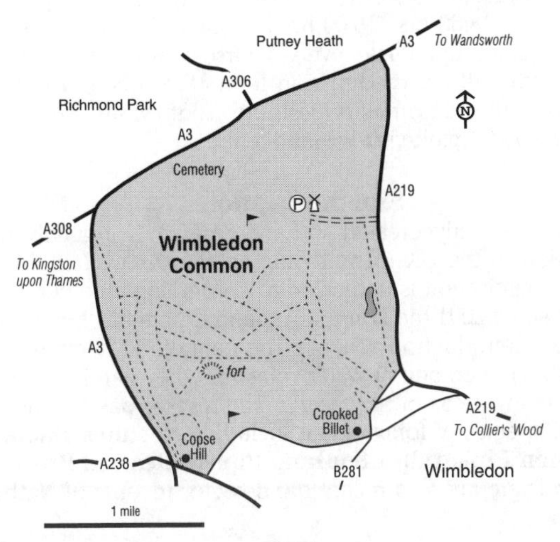

Site description

All the target species may be seen hunting over Wimbledon Common during the summer. The small population of **Daubenton's Bats** are best looked for hunting over Queensmere and Kingsmere. The common is large and incorporates sports facilities as well as a diverse mix of natural habitats such as heathland, woodland, scrub and ponds. Bat walks on Wimbledon Common are led by the London Bat Group: www.londonbats.org.uk. Other species present on the common include **Weasel**, **Stoat**, **Red Fox**, **Bank Vole**, **Field Vole**, **Mole** and **Hedgehog**, and **Water, Common** and **Pygmy Shrews**.

Other information

Noctules may also be found feeding in most of the capital's parks including Hyde Park, Regent's Park, Oxleas Wood and Hampstead Heath.

20 London Wetland Centre, Barnes

Grid ref: TQ226768

Key species

Water Vole, bats.

Access

The London Wetland Centre, a reserve of the Wildfowl and Wetlands Trust, is easily accessible by car, being situated less than one mile (1.6km) from the South Circular (A205) at Roehampton and from the A4 at Hammersmith. There is ample parking available. The routes around the reserve are wheelchair accessible as are most of the hides. Admission charges for WWT non-members are: adults £7.25, children £4.50, family ticket £18.50, concessions £6. The reserve is open daily (except Christmas Day) from 09:30–18:00 hrs (16:00 hrs in winter). Bat watchers will welcome that it stays open late every Thursday from 25 May until 21 September, with half price admission from 18:00 hrs and last entry at 20:00 hrs. For further information visit the Wildfowl and Wetlands Trust website: www.wwt.org.uk/visit/wetlandcentre/

Site description

This relatively recently created wetland reserve is arguably the best urban location in the UK for watching wildlife. As well as the diverse birdlife, the organisation is dedicated to conserving all other aspects of wildlife and since 2001 the London Wetland Centre has been running a **Water Vole** reintroduction scheme. The centre's 40 hectares are now home to a thriving colony of Water Voles, and they can be viewed with patience in many areas of the reserve. The reserve has also attracted a total of nine species of foraging bat including **Noctules**, **Daubenton's Bats**, **Common Pipistrelles**, **Soprano Pipistrelles** and **Brown Long-eared Bats**. There are also recent bat detector records of **Nathusius's Pipistrelles**.

21 Beddington Park, London **Grid ref: TQ291654**

Key species

Serotine, Noctule, Common Pipistrelle, Soprano Pipistrelle.

Access

Beddington is between Sutton and Croydon. Beddington Park can be accessed via London Road (A237) to the west or Croydon Road (A232) to the south.

Site description

Beddington Park was originally part of the deer park attached to Carew Manor which at one time occupied almost all the land between Mitcham Common, Beddington Lane, Croydon Road and London Road.

Small numbers of **Serotines** and **Noctules** may be encountered hunting over the park during summer, whilst **pipistrelles** are relatively common throughout the area.

22 Wanstead Flats, London **Grid ref: TQ409864**

Key species

Leisler's Bat, Noctule, Daubenton's Bat.

Access

This parkland area is sandwiched between Aldersbrook Road (A116) and Forest Drive (A114). There are car parks on both of these roads.

Site description

Wanstead Flats is the southernmost part of Epping Forest and excellent for bats. It is an area of rough grassland with mown areas and several ponds. Small numbers of **Leisler's Bats** may be seen hunting at this site during the summer. It is worth checking any white street lamps in the area after dark, as bats are often attracted to this artificial light source in search of insects. **Noctules** are also fairly easy to observe here as the lack of trees makes viewing this high-flying species fairly straightforward. A few **Daubenton's Bats** hunt over the pools, and **pipistrelles** also occur here.

23 Berwick Ponds, London **Grid ref: TQ540835**

Key species

Noctule, Serotine, Common Pipistrelle, Soprano Pipistrelle.

Access

Located in east London, just north-east of Rainham, Berwick Ponds is accessed via Berwick Pond Road at the above grid reference.

Site description

During the summer, up to 30 **Noctules** may be seen hunting over the reed-fringed ponds, a fantastic place to watch this species. Small numbers of **Serotines** also feed over the ponds and can usually be located amongst the larger numbers of **pipistrelles**.

24 Elmley Marshes RSPB Reserve, Kent

Grid ref: TQ965674

Key species

Water Vole.

Access

Located on the Isle of Sheppey in north Kent, Elmley Marshes are signposted off the A249, about one mile (1.6km) beyond the Kingsferry Bridge.

Site description

This RSPB reserve is one of the finest examples of coastal grazing marsh in the country. The reserve has a significant population of **Water Voles** and has been included in the 'Water Vole Key Sites' programme, a project partly funded by the People's Trust for Endangered Species (PTES). This species benefits from the extensive cover of Sea Club-rush that grows along the edges of ditches, channels and pools, providing food and cover throughout the year. Other species on the reserve include **Harvest Mouse**, **Brown Hare**, **Water Shrew**, **Pygmy Shrew**, **Common Shrew**, **Field Vole**, **Red Fox**, **Stoat**, **Weasel** and **American Mink**.

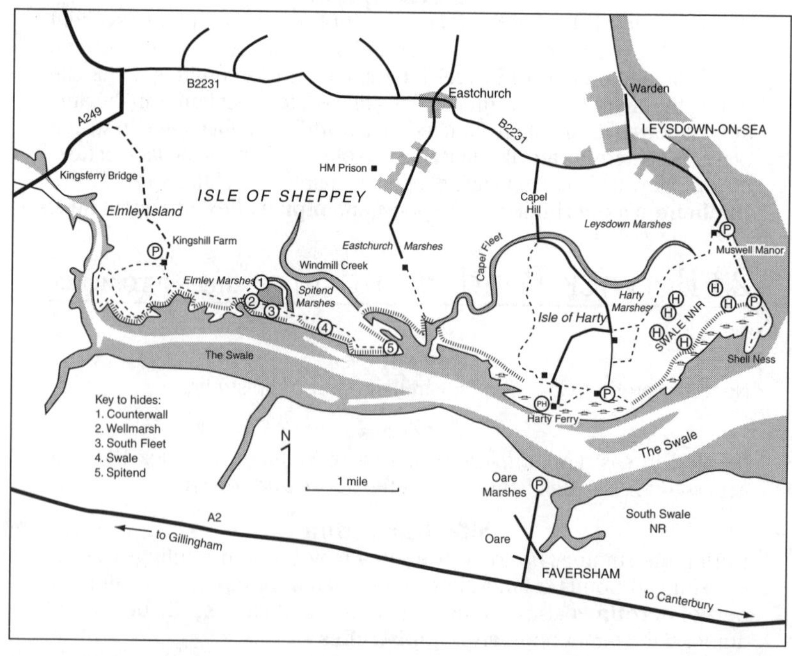

25 Herne Bay, Kent

Grid ref: TR176684

Key species

Grey Seal, Common Seal.

Access

The town of Herne Bay is on the north coast of Kent. The sand flats and beach are easily accessed through the town.

Site description

Both **Grey Seals** (in small numbers) and **Common Seals** may be seen hauled out along the Barrow Sands sandbank at Herne Bay. The Mammals Trust UK runs a trip each year to watch the seals at this location.

26 Stodmarsh NNR, Kent

Grid ref: TR221609

Key species

Water Vole, Water Shrew, American Mink.

Access

Stodmarsh is in the Stour Valley between Stodmarsh village and Grove Ferry, about 500m from the A28 Canterbury to Thanet road. To reach the main entrance and car park take the track beside the small village green, next to the Red Lion Inn, and follow the reserve signs. The track is 500m long and suitable for vehicles but uneven in places with loose material. The Grove Ferry entrance is from the roadside. Toilets are available in the car park, which is free. There is open access to the reserve at all times via the network of nature trails and public footpaths, although certain parts are designated as sanctuary areas where access is not permitted.

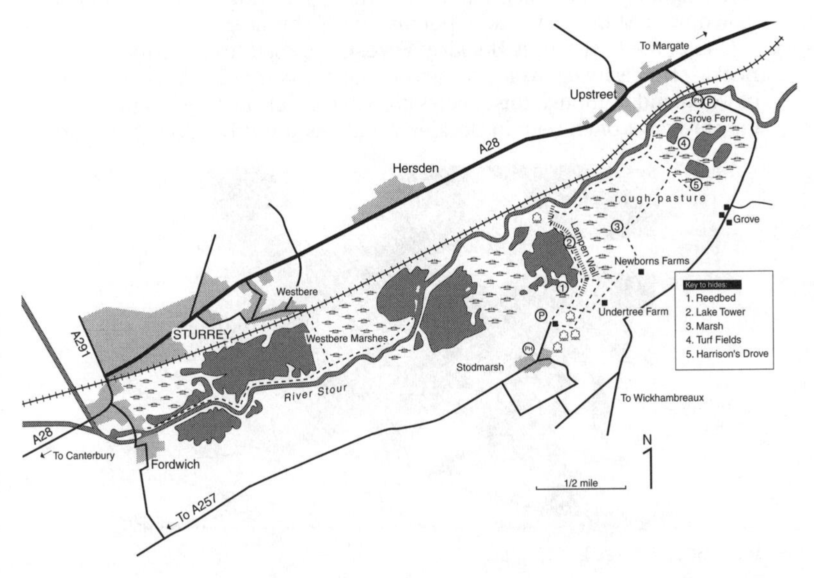

Site description

This 163-hectare wetland nature reserve incorporates reedbeds, shallow lagoons, grazing meadows and wet woodlands. **Water Vole** populations at Stodmarsh are doing well: indeed, they are starting to colonise new areas within the reserve. Their populations are regularly surveyed by live trapping and by checking for signs, such as footprints and droppings.

There are four bird hides and an observation mound, all of which overlook suitable Water Vole habitat, on the easy access trail. **Water Shrews** may be found in similar habitat throughout the reserve and patience is often rewarded from the viewing points. Other mammals that may be encountered include **American Mink**, **Red Foxes** and **Stoats**, whilst **Otters** occasionally pass through.

27 Kent/East Sussex border

Key species

Wild Boar.

Wild Boar are present in a number of parishes in south-east England. These include:

Kent Aldington, Appledore, Bilsington, Lympne, Ruckinge, Stone-cum-Ebony, Warehorne, Wittersham, Woodchurch
East Sussex Beckley, Kenardington, Peasmarsh, Udimore
(Source – Central Science Laboratory report via www.britishwildboar.org.uk website).

The parishes listed are in an area bordered by the towns of Ashford, Kent, and Heathfield, East Sussex. The area consists of a network of mixed woodlands and agricultural fields, thus providing ideal habitat. The population in this area is unknown but may be as many as 400 individuals. Other sightings have been reported north to Tonbridge, and the animal may now exist in a network of populations in this area.

One of the best sites is **Beckley Forest**, accessed from the road from Beckley to Peasmarsh which passes through the centre of a large expanse of woodland. Although this species probably occurs throughout the area, they are most often seen in Beckley Woods. As you drive along this road

Wild Boar (*V. Ree*)

you will see several different woods and reserves. At the woodland signposted Beckley Woods, walk through the barrier and bear to your left. You will go down into a dip after about 400m and up the other side. Examine the various rides in the area looking for signs such as diggings and damaged fences. The boar are most active at dawn and dusk.

Please remember that sows with piglets can be extremely dangerous, so take care when visiting this site.

28 Rye Harbour Nature Reserve, East Sussex

Grid ref: TQ940187

Key species

American Mink, Water Vole.

Access

Rye Harbour can be accessed on a minor road south of Rye off the A259. There is a large car park at Rye Harbour village. The reserve is flat and many of the hides have easy access, although certain sections of footpath require you to climb over stiles.

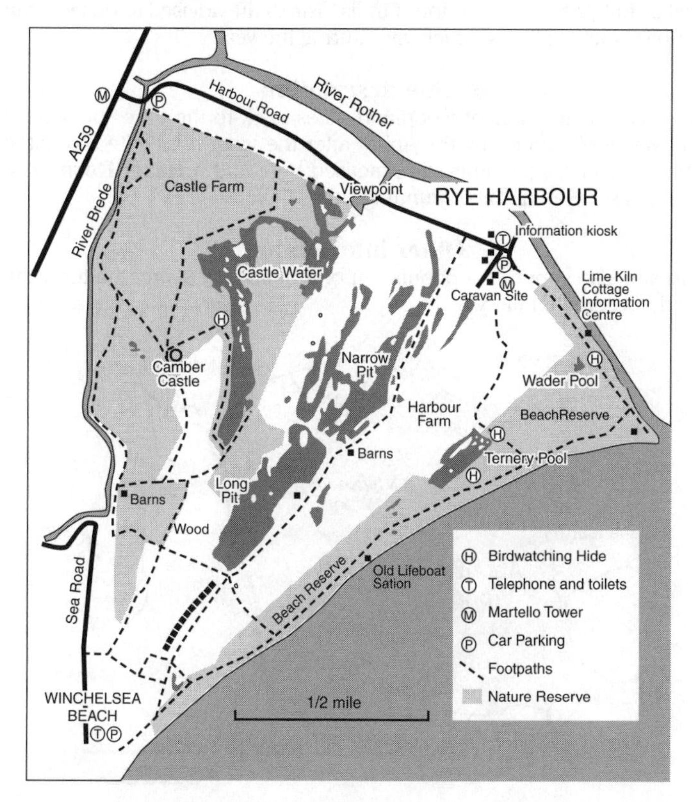

Site description

In 1970 334 hectares of Rye Harbour were declared as a Local Nature Reserve, parts of which are now managed by the Sussex Wildlife Trust. The habitats that make up the reserve are quite varied with extensive areas of vegetated shingle, grassland and open water. **American Mink** occur in good numbers despite trapping efforts by the countryside management team. They can be seen anywhere on the reserve but sightings are extremely scarce from Ternery Pool. Castle Water is a good spot as this is where the greatest numbers are trapped. **Water Voles** remain relatively common at Castle Farm and are occasionally seen elsewhere on the reserve.

A full site list of mammals (and other vertebrates) is at www.yates.clara.net/vertlist.html.

29 Abbot's Wood, East Sussex **Grid ref: TQ555073**

Key species

Hazel Dormouse.

Access

This site is signposted off the A22 between Hailsham and Polegate, and off the A27 at Wilmington. There is a pay and display car park on site, as well as toilets and marked forest trails. Visitors are advised to observe the car park locking times, which vary during the year.

Site description

This 360-hectare ancient woodland dates back to the times of Henry I and was looked after by the Abbot after the wood was gifted to Battle Abbey. Hazel coppicing is still practised here and a **Hazel Dormouse** nest-box scheme has been running since 1995.

Other information

This site also supports a population of a nationally scarce butterfly, the Pearl-bordered Fritillary.

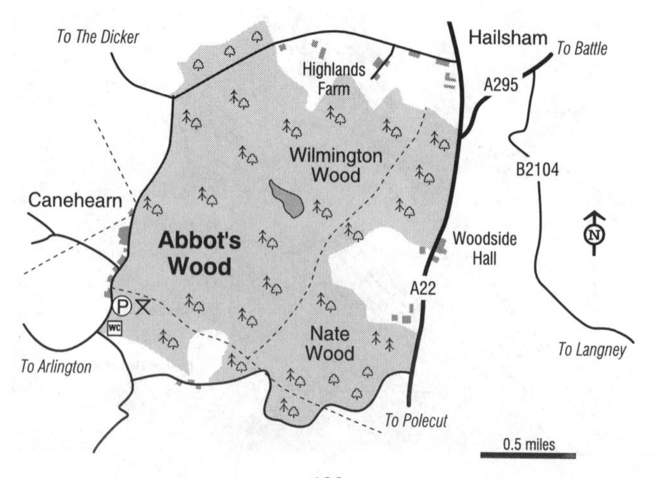

EAST ANGLIA

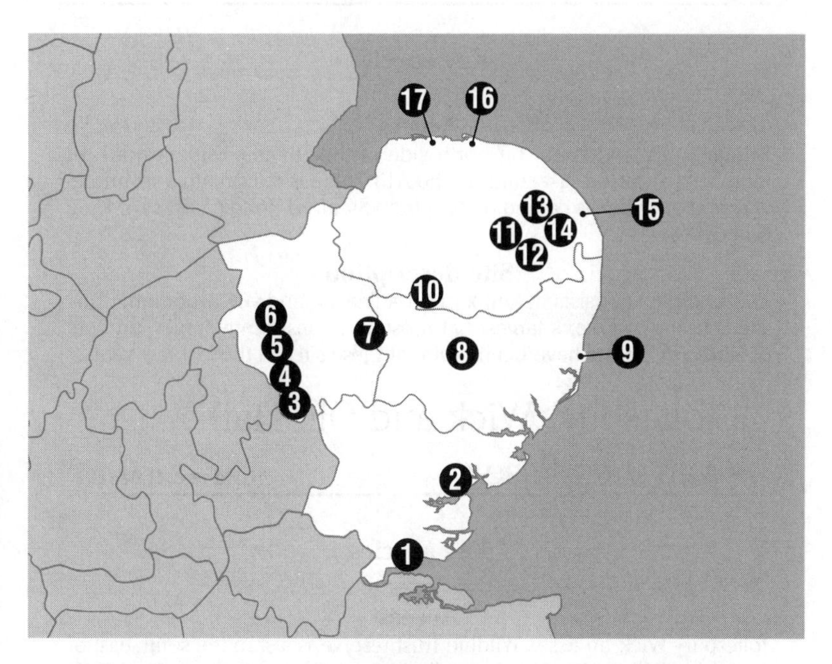

1 Hangman's Wood and Deneholes
2 Tollesbury Wick and Old Hall Marshes
3 Fowlmere RSPB Reserve
4 Wimpole Hall
5 Eversden Wood
6 Woodwalton Fen, Holme Fen and Monks Wood
7 Wicken Fen NT
8 Bradfield Woods National Nature Reserve
9 Minsmere RSPB Reserve
10 Breckland/ Thetford Forest
11 Cringleford Marsh, Norwich
12 Strumpshaw Fen RSPB Reserve
13 Barton Broad
14 Hickling Broad and Stubb Mill
15 Horsey and Winterton-on-Sea
16 Blakeney Point and Cley NWT Reserve
17 Holkham

1 Hangman's Wood and Deneholes, Essex

Grid ref: TQ630793

Key species

Natterer's Bat.

Access

Hangman's Wood is on the north side of the Thames Estuary, north of Little Thurrock and just south of the A13. There is a footpath and bridleway and entrance is gained on foot from Stanford Road (A1013) or King Edward Drive.

Site description

This complex consists of a mix of broadleaved and yew woodland. This site is home to Essex's largest bat roost and consistently large numbers of **Natterer's Bats** have been found at these sites in the last few years.

2 Tollesbury Wick and Old Hall Marshes, Essex

Grid ref: TL975101

Key species

Water Vole.

Access

Tollesbury Wick, an Essex Wildlife Trust reserve, is just to the south-east of Tollesbury. Follow the B1023 to Tollesbury via Tiptree, leaving the A12 at Kelvedon, then follow Woodrolfe Road towards the marina and car park at Woodrolfe Green. The reserve is accessible at all times along the seawall.

Site description

This freshwater grazing marsh is a Site of Special Scientific Interest (SSSI) where traditional farming methods are employed to be sympathetic to wildlife. Good numbers of **Water Voles** occur in the reed beds. **Field Voles** and **Pygmy Shrews** also occur.

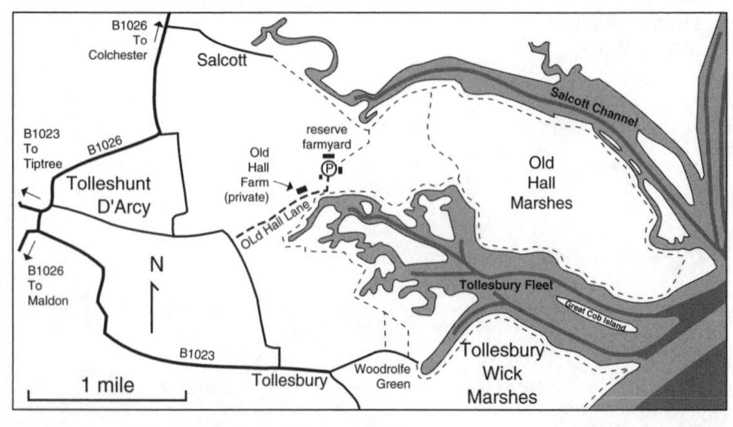

Other information

Sheep ticks can be a problem here between April and June when it is advisable to wear long trousers and to keep out of long vegetation.

3 Fowlmere RSPB Reserve, Cambridgeshire

Grid ref: TL408458

Key species

Otter, Water Shrew.

Access

The reserve is seven miles (11km) south of Cambridge between Fowlmere and Melbourn and is signposted off the A10 between Royston and Cambridge.

Site description

This 40-hectare wetland reserve can be accessed via a series of boardwalks and pathways. There are three hides on site, one of which is wheelchair accessible.

Otter: Fowlmere's reedbeds and pools are fed by natural chalk springs and provide ideal habitat for Otters. In the past five years this species has started to be seen regularly on the reserve, and a family was present in the summer of 2004. The best area is the Reedbed Hide although they are also occasionally seen from the Spring Hide. Warm, still evenings are best and sightings often peak in late June and early July, during which time the Otters appear to hunt wildfowl.

American Mink: Mink used to be seen regularly around the reserve but sightings have become less frequent since the return of Otters to the area. September, when the young become active, is often the best month, and in 2005 there were several sightings during this month.

Water Shrew: The reserve has a large population of Water Shrews and they are often seen or heard from the hides or paths that run along the streams in the reserve. April and May are particularly good times to see the shrews, as they are often in family parties of 5–10 individuals.

Water Shrew (*P. Twisk*)

Other species

Fallow Deer are regularly seen in the meadow in the south-east section of the reserve and **Muntjac** occur throughout. There is an active **Badger** sett on the reserve and **Brown Hares** occur in the surrounding farmland. **Common Pipistrelles** are frequently encountered between the Reedbed and Spring Hides. **Noctules** are occasionally seen over the reedbed although sightings are becoming increasingly rare.

4 Wimpole Hall, Cambridgeshire

Grid ref: TL335510

Key species

Serotine, Barbastelle, Natterer's Bat, Brown Long-eared Bat.

Access

This National Trust property is approximately 15 miles (24km) west of Cambridge and is signposted off the A1198 and A603. The parkland is open all year and a public footpath follows the road through the park. The woods can be entered from the minor road that runs north from the B1042 towards Wimpole Hall. Continue past the entrance to the Hall and up the hill. Park carefully where the road makes a right-angle turn to the left at the top of the hill. Parking space is very limited, and it may be better to use the car park at Wimpole Hall and walk up the hill. A public footpath heads due east from the corner of the road and runs alongside the woods. There is also a permissive path running through the wood a little below the corner. Bats can be seen and heard a short distance along either path.

Site description

The cattle-grazed pastureland surrounding Wimpole Hall provides ideal hunting habitat for the **Serotines** that take advantage of the profusion of large insects, such as dung beetles. A relatively narrow strip of secondary woodland around the edge of the estate, known as The Belts, together with the privately owned Eversden Wood, are designated a Special Area of Conservation because of the presence of roosting Barbastelles. The **Barbastelles** can usually be heard and sometimes seen flying along the paths both inside and along the northern edge of the wood. They are active shortly after sunset but typically move away from the woods after half an hour or so. Both of the 'common' **pipistrelle** species are present in the wood, and **Brown Long-Eared Bats** and **Natterer's Bats** are sometimes heard. **Noctules** are occasionally encountered.

5 Eversden Wood, Cambridgeshire

Grid ref: TL.340.526

Key species

Barbastelle.

Access

The wood is situated seven miles (11km) west of Cambridge. From the A603 turn towards Wimpole village, continue for approximately two miles (3km) and park somewhere near the sharp left-hand turn in the road. From here there are two public footpaths that lead north-west to skirt either edge of Eversden Wood. Public footpaths also run through the wood.

Site description

Eversden Wood is a relatively isolated 66-hectare site. **Barbastelles** were first located in woodlands in the area in 2001.

6 Woodwalton Fen, Holme Fen and Monks Wood, Cambridgeshire

Key species

Chinese Water Deer.

Access

Woodwalton Fen (TL220846) is four miles (6km) west of Ramsey on Chapel Road, Ramsey Heights. There is off-road parking available at the site and hides for viewing. Holme Fen (TL223886) is a mile (1.6km) east of the village of Holme, and although there are no car parking facilities on-site there are lay-bys on the nearby road. Similar parking arrangements occur at Monks Wood with a lay-by on the B1090 just 300m from the reserve. Monks Wood (TL199801) is two thirds of a mile (1km) west of Woodwalton.

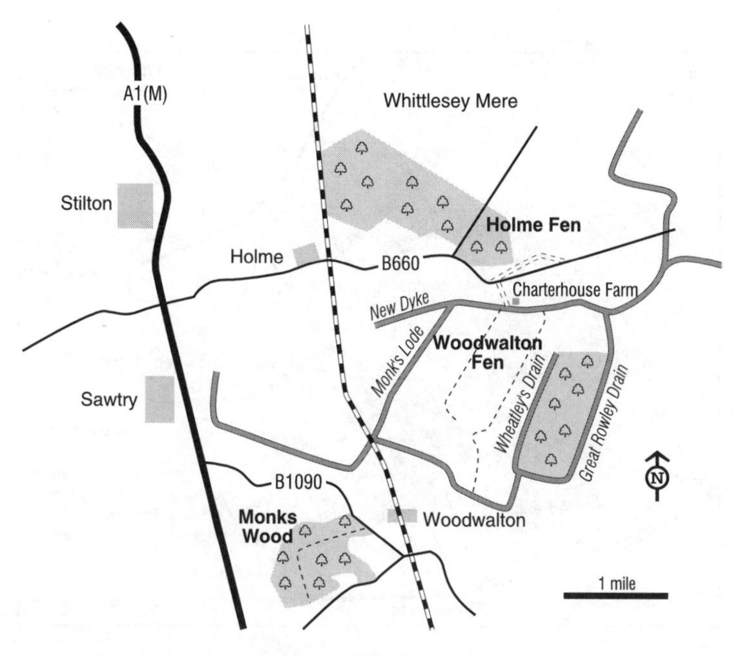

Site description

This area of Cambridgeshire provides ideal habitat for **Chinese Water Deer** and the population is thriving. There are estimated to be at least 200 deer in the area. They may be observed at any of the National Nature Reserves listed, with Woodwalton and Holme Fens probably providing the best opportunity. To give yourself the best chance, watch over suitable habitat at dawn or dusk.

7 Wicken Fen NT, Cambridgeshire

Grid ref: TL563704

Key species

Water Shrew, Water Vole.

Access

Wicken Fen is five miles (8km) south-west of Soham off the A1123. Follow signposts from Wicken village. There is a small charge for non National Trust members.

Site description

This is Britain's oldest nature reserve and has been in existence for over 100 years.

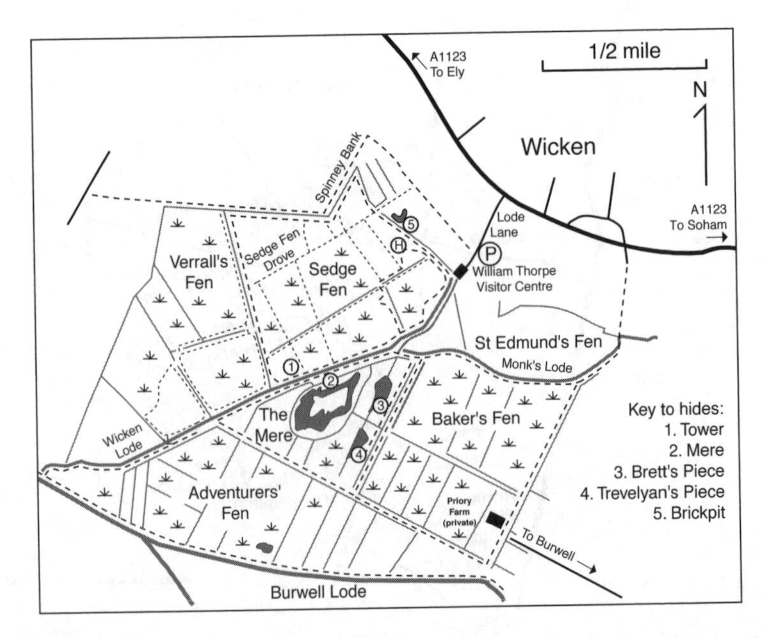

Water Shrews are commonly found dead around the reserve, although live animals are more elusive. The managed sedge field is a good place to look during summer and they are also present in the fen–carr areas. Other small mammals are plentiful with large populations of **Harvest Mice** in the sedge fields and on Adventurer's Fen; and **Water Voles** are common along the lodes and ditches. **Wood Mice** are common in the fen carr, sedge and litter fields. **Bank Voles, Field Voles, Pygmy Shrews** and **Common Shrews** can also be seen, and **Brown Hares** breed on Adventurer's Fen.

8 Bradfield Woods National Nature Reserve, Suffolk

Grid ref: TL934579

Key species

Muntjac, Roe Deer, Hazel Dormouse, Yellow-necked Mouse.

Access

This site is approximately 13 miles (21km) south-east of Bury St Edmunds and eight miles (13km) west of Stowmarket, on the road between the villages of Bradfield St George and Felsham. There is a car park at the reserve entrance and three nature trails through the woods, although these become impassable for disabled visitors during the winter. Toilets are available at the entrance.

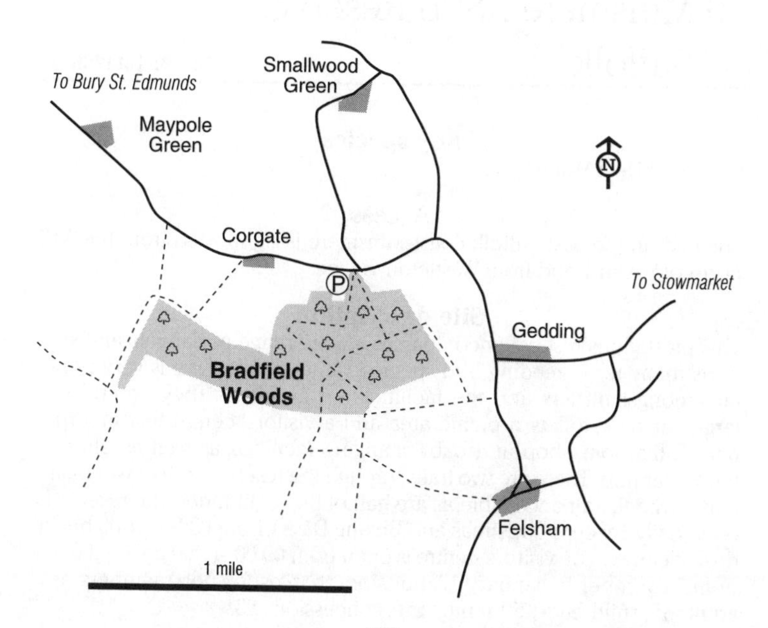

Site description

This is one of the finest ancient woodlands in East Anglia and continuous coppicing dates back to 1252 AD. The woods, which cover 64 hectares, are owned by the Suffolk Wildlife Trust and managed in conjunction with Natural England.

Muntjac: There is a large population of Muntjac within the woods and they cause a few headaches for Natural England as they browse the regrowth of the hazel stools. This species can often be seen along the woodland rides early in the morning.

Hazel Dormouse: This classic coppice species is uncommon across most of East Anglia; the woods at Bradfield contain one of the largest populations within the region, although viewing them is obviously extremely difficult. The best areas are the more mature stands of Hazel coppice where there is a profusion of other fruiting tree and shrub species in close proximity.

Yellow-necked Mouse: Like the previous species the Yellow-necked Mouse is rare within the region but the ancient woods at Bradfield contain a large population.

Other mammal species: As well as Muntjac, Bradfield Woods supports a large population of **Roe Deer**, whilst **Red Deer** and **Fallow Deer** are infrequent visitors. **Badgers** are resident.

Other information

Over 250 vascular plant species have been recorded within the woodlands. Notable species include the nationally rare Oxlip.

9 Minsmere RSPB Reserve, Suffolk

Grid ref: TM470671

Key species

Otter, Red Deer, Water Vole.

Access

Located on the east Suffolk coast, Minsmere is signposted from the A12 north of Yoxford, and from Westleton village.

Site description

This prestigious reserve encompasses a wide range of habitats and supports many rare breeding bird species. Hence the reserve is very popular amongst birders and the facilities are excellent. They comprise a large car park, toilets, a picnic area and a visitors' centre that incorporates a tearoom, shop and baby-changing facilities, as well as offering binocular hire. There are two trails through the reserve, both of which are partly wheelchair accessible, as are half of the eight hides. The reserve is open daily (except Christmas and Boxing Days), from 09:00–21:00 hrs or dusk if earlier. The visitors' centre is open from 09:00–17:00 hrs (16:00 hrs from November to January). Admission charges for non-members are: adults £5, children £1.50, family £10, concessions £3.

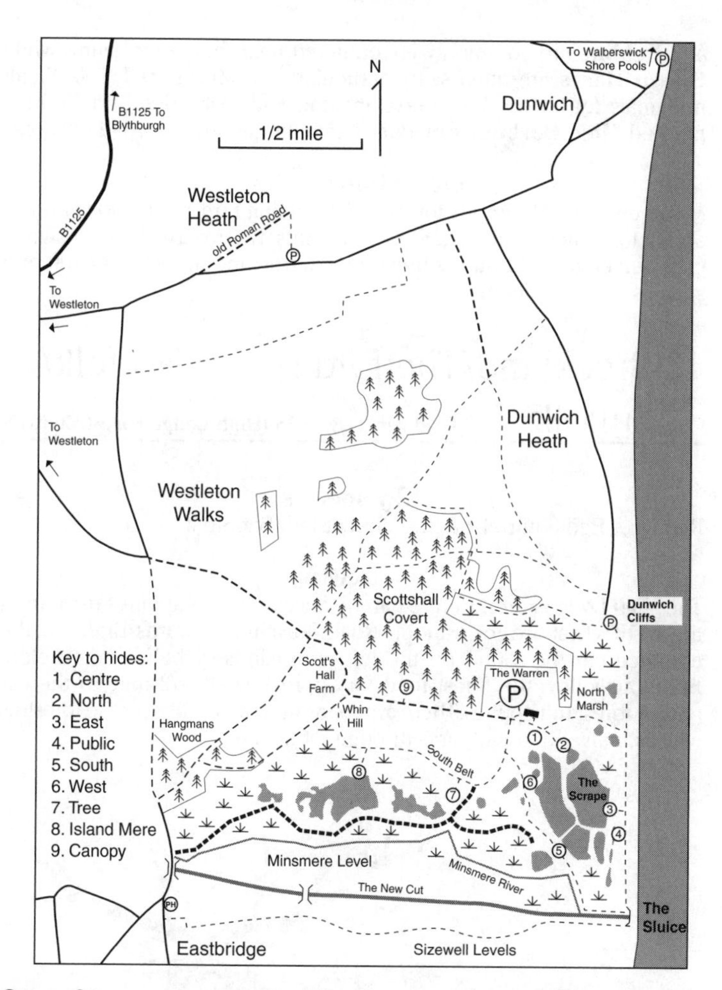

Otter: Otters are now seen at Minsmere throughout the year. Although they can be encountered at any time of day they are perhaps easiest to see in the late evening. They are regularly seen around the Island Mere particularly in late April and May. Warm, still evenings are best, particularly after 19:00 hrs.

Red Deer: Red Deer can be found throughout the year, and the RSPB offers trips to see this species during the rutting season. They can often be seen at dusk along the road from Scott's Hall Farm to Westleton and they are also regularly seen on the heathland areas of the reserve.

Water Vole: Water Voles, increasingly difficult to see elsewhere, are regularly seen on the reserve, particularly from Island Mere and Tree Hides.

Other mammals: Minsmere provides a home to a large number of other mammals. **Red Foxes** are often seen hunting the rabbit warren behind the Island Mere Hide. **Stoats** and **Weasels** are occasionally encountered on trails throughout the reserve. **Muntjac**, **Grey Squirrels**

and **Rabbits** are commonly encountered near the visitor centre while **Brown Hares** are often seen, particularly on Minsmere Levels. Small mammals found on the reserve include **Harvest Mice** and **Yellow-necked Mice**. **Harbour Porpoises** are infrequently recorded offshore.

Other information

Minsmere is most famous for its rare breeding bird populations and its ability to attract migrants and rare vagrants. The reserve holds a significant percentage of Britain's breeding Bitterns, and in total over 100 bird species breed on the reserve.

10 Breckland/Thetford Forest, Norfolk/ Suffolk

Grid ref: TL807855 (High Lodge Forest Centre)

Key species

Red Deer, Red Squirrel, Barbastelle, Leisler's Bat, Stoat.

Access

The High Lodge Forest Centre can be accessed via the forest drive from the B1107 Thetford to Brandon Road. Travelling towards Brandon, the entrance can be found on the left, approximately half a mile before Brandon. Follow the forest road for approximately 1/2 mile to the car park. A forest toll applies. There are roads and tracks all over the forest, as well as many other car parks throughout the area.

Red Deer (*James Gilroy*)

Site description

Thetford Forest is an extensive area of coniferous plantations and sandy heath that straddles the border of Norfolk and Suffolk. Indeed, its 2,000 hectares make up the largest lowland pine forest in the UK. The area is underlain by a mixture of chalk and sandy soils that were traditionally cultivated for short periods before being left fallow. The term 'breck' refers to such broken ground that is used temporarily for cultivation. The open, disturbed landscape coupled with its continental-like climate has resulted in a unique ecological community.

Red Deer: This vast area of pine forests and grass heath is home to good numbers of Red Deer. In 1992 a herd of about 100 deer could be encountered on Forestry Commission land in the area. Herds are often difficult to come by and it is more likely that the potential watcher will come across lone individuals or small groups of this species crossing tracks or other open areas within the forest. **Roe Deer** are also common throughout Breckland and are often seen in fields from the road. **Muntjac** are also numerous.

Barbastelle: The Forestry Commission has established a bat hibernaculum at High Lodge. This has attracted ten species of bat in total, six of which regularly spend part of the winter hibernating inside. It has proved to be a very important site for Barbastelles. Possibly the best time to locate them at the site is during swarming activity in the autumn as they prepare to hibernate. However it would be worth checking areas in the forest during the summer as well to try to determine their distribution in the forest.

Leisler's Bat: Small numbers of Leisler's Bats are regularly caught during monitoring of nest boxes within Thetford Forest. The Norfolk Bat Group runs bat box sessions from April to October, and casual observers are welcome to join them. Up to six species are seen regularly on these sessions including **Brown Long-eared Bats**, **Common** and **Soprano Pipistrelles**, **Noctules** and **Leisler's Bats**.

Red Squirrel: East Anglia's last remnant population of Red Squirrels maintains a toehold in Breckland's vast swathes of coniferous forest where they coexist with Grey Squirrels. In autumn 2003 the population was thought to be between only 50 and 100 animals, although there had been no confirmed sightings between 2001 and 2003. At this time, as part of a Species Recovery Programme, a reintroduction scheme and a captive-breeding programme had to be stopped due to the appearance of the parapox virus in the area. The programme has since been restarted and small numbers of Red Squirrels can be found in various parts of the forest. Sites worth checking include Rishbeth Wood (TL841841) by the Thetford Warren Lodge and the woodland surrounding the High Lodge Forest Centre (TL813852). It would be wise to seek the latest information before visiting to look for this species. The High Lodge Forest visitor centre, which is open at weekends, can be contacted on 01842 815434.

Stoat: The Norfolk Wildlife Trust reserve of Weeting Heath (TL757878) offers short grassland habitat favoured by the rare Stone Curlew and, more relevantly, **Rabbits**. Warrens riddle the turf to either side of the road making ideal hunting terrain for **Stoats**. I have personally observed

Stoats throughout the year at this site but I believe the best time to watch is during the winter. At this time, more individuals may be observed and they will often be displaying differing degrees of winter coloration (ermine). The reserve is on the minor road between Weeting and Hockwold-cum-Wilton, north-west of Thetford.

11 Cringleford Marsh, Norwich, Norfolk

Grid ref: TG197068

Key species

Noctule, Water Vole.

Access

Situated on the south-west side of Norwich, between the University of East Anglia and Cringleford village, Cringleford Marsh is easily accessed via the Yare Valley walk from either the north or the south. The best point of access is from the Yare Valley walk car park at the junction of North Park Avenue and Bluebell Road at TG194074. From here walk into the university grounds and then turn south along the footpath alongside the eastern end of the Broad. Shortly after turning along the southern edge of the Broad turn left onto the boardwalk that takes you south through Cringleford Marsh. Alternatively, this walk can be accessed from Cringleford village just south of the A11. Although the university Broad, Cringleford Marsh and surrounding grasslands are open to the public the university campus itself should not be entered.

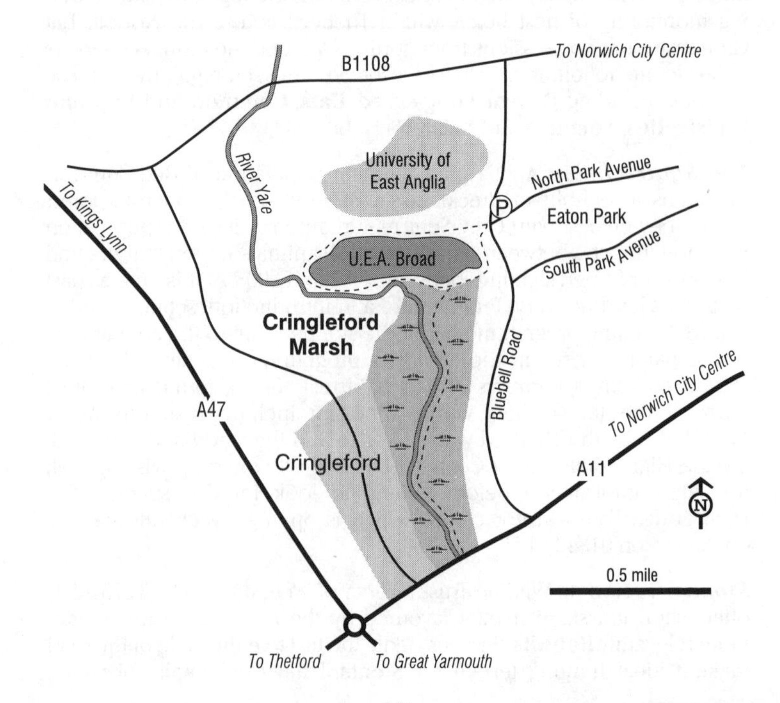

Site description

Cringleford Marsh consists of areas of reedbed along with other typical marshy habitats and is managed by staff from the university who carry out periodic burning of the Common Reed.

Noctule: During the summer good numbers of Noctules can be watched for long periods before complete darkness falls hunting over the University Broad, the surrounding grassland or Cringleford Marsh.

Water Vole: The marsh is one of the best sites in the region to see Water Voles. Although the single point of access is from the Yare Valley boardwalk, which is often busy at weekends and other times, it is possible to observe this species early in the morning or at other quieter times

12 Strumpshaw Fen RSPB Reserve, Norfolk

Grid ref: TG340067

Key species

Otter, Water Vole, Chinese Water Deer.

Access

Strumpshaw Fen is signposted from Brundall, off the A47 between Norwich and Great Yarmouth. The reserve is open from dawn until dusk daily (except Christmas Day). There is a car park, disabled toilets and a visitors' centre.

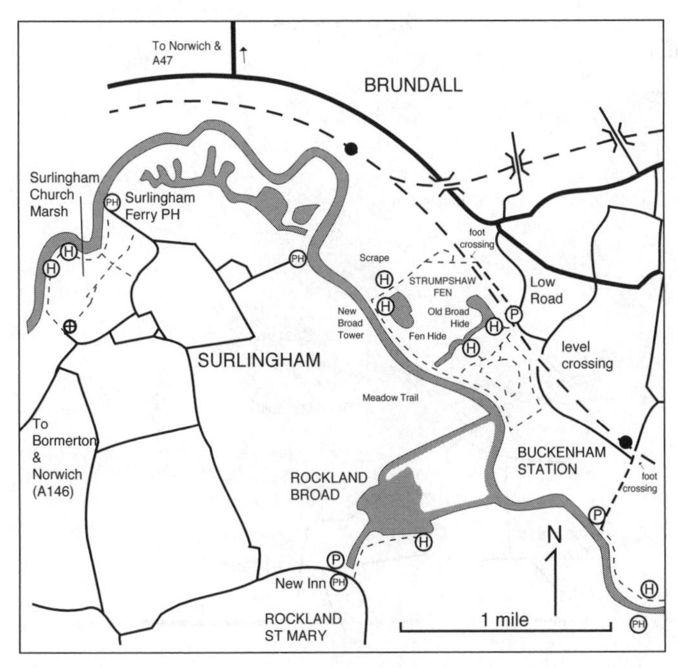

Site description

Strumpshaw Fen is situated in the heart of the Norfolk Broads and is dominated by a large reedbed adjacent to the River Yare. There is a visitor centre, several hides and a circular walk around the reserve.

Water Vole populations within the Norfolk Broads still appear to be fairly healthy and the RSPB reserve at Strumpshaw provides one of the best opportunities to observe this declining animal. There are three hides that with patience may produce the goods, the best probably being the New Broad Tower, adjacent to the River Yare. Other small mammals are common with **Harvest Mice** breeding in the long grass and reedbed and **Water Shrews** on the pools. **Brown Long-eared Bats** and **pipistrelles** hunt over the reserve at twilight.

Otters have started to be observed fairly regularly from the Brick Hide visitor centre: late evenings during spring are the best time. **Chinese Water Deer** breed on the reserve and with luck can be seen throughout the year but are best looked for on the marshy area on the southern side of the reserve.

Other information

This reserve has a large population of Swallowtail butterflies.

13 Barton Broad, Norfolk

Grid ref: TG361215

Key species

Otter.

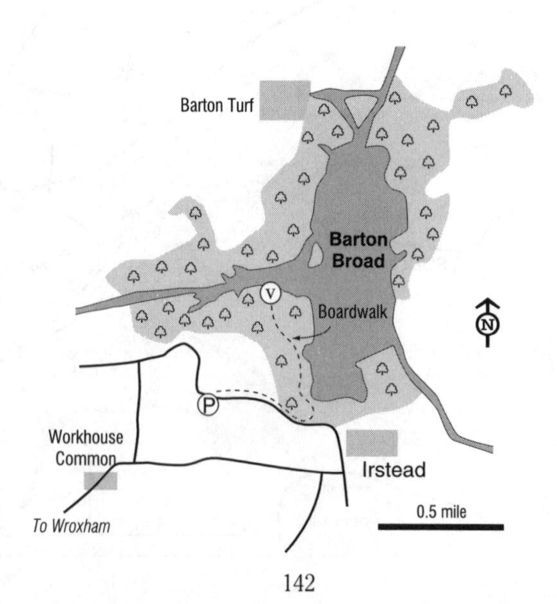

Access

Barton Broad is approximately 12 miles (19km) north-west of Norwich, just west of the village of Irstead. There are two car parks, a disabled-only area by the start of the boardwalk and further parking just east of here by the Barton Angler Country Inn. The recently built boardwalk meanders through alder and birch carr to the Broad's southern edge, where a viewpoint allows observation of most of the Broad. The boardwalk is wheelchair accessible.

Site description

Norfolk's second largest Broad is owned by the Norfolk Wildlife Trust and managed in partnership with Natural England, with the whole area designated a National Nature Reserve. The Broad itself is surrounded by carr woodland, fen, reedbed and marshland.

Otter: The successful recolonisation by Otters has been brought about partly due to work that has been carried out in restoring fish populations and improving the water quality, as well as the general upturn in fortunes for the Otter itself. The viewpoint from the boardwalk at the southern end of the Broad is the only place from which casual visitors can watch. Otters are best looked for during calm summer evenings. Local fishermen have reported seeing up to six individuals at once which indicates successful breeding in the general area.

14 Hickling Broad and Stubb Mill, Norfolk

Key species

Chinese Water Deer.

Site description

Over 300 **Chinese Water Deer** roam freely in the Norfolk Broads and the population of the area looks set to flourish. As may be expected, they are most easily observed when they feed away from the marshes and reedbeds and come into the neighbouring farmland and woodland areas.

Key areas

Hickling Broad NWT, NNR (TG420204). This is probably one of the most reliable sites in the Broads. Park at the Potter Heigham Church and follow the Weaver's Way footpath to the Broad. You have to walk through farmland and woodland before you reach the open water, and individuals can often be encountered foraging in the arable fields close to the woodland edge. Red Deer are also sometimes seen here.

Stubb Mill (TG430220). The grazing marshes around Stubb Mill often attract grazing Chinese Water Deer. The prime time is dusk when success, particularly during winter, should be guaranteed. From Hickling village, take Stubb Road to the Norfolk Wildlife Trust visitor centre and park in the car park. Then walk back to Stubb Road, turn right and walk down about two thirds of a mile (1km) to the abandoned mill. Increasing numbers are also using the fields along Stubb Road itself as feeding areas.

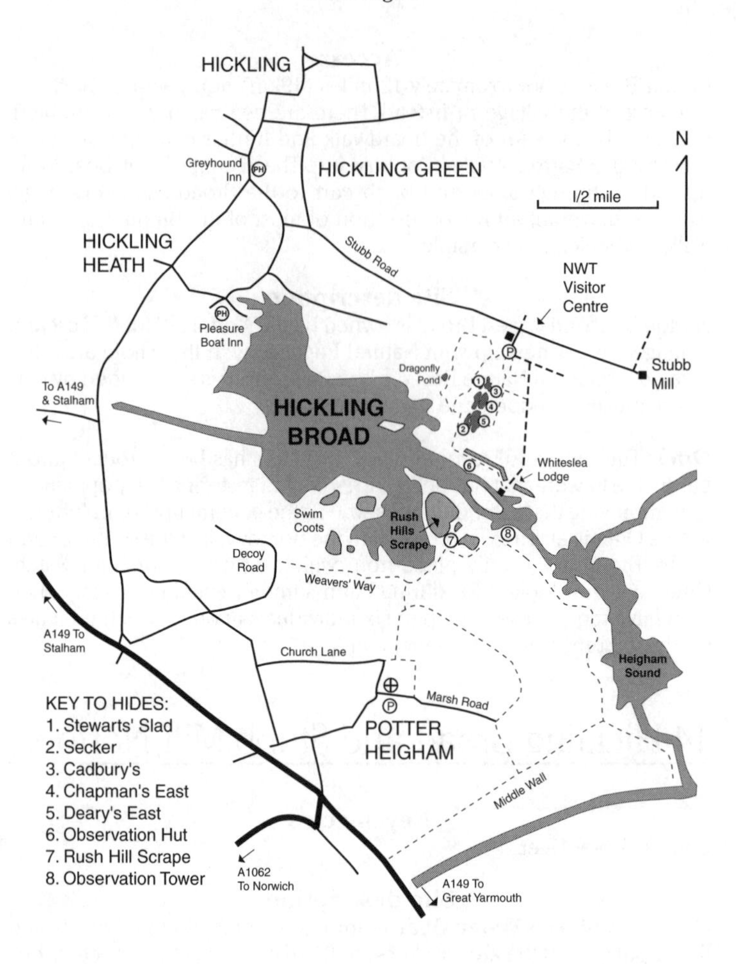

Chinese Water Deer (*P. Barrett*)

15 Horsey and Winterton-on-Sea, Norfolk

Grid ref: TG480227

Key species

Grey Seal, Common Seal, Harbour Porpoise.

Access

This fairly extensive area of coastal east Norfolk has several access points. At the southern end lies the village of Winterton-on-Sea where there is a beach car park. From here it is possible to walk north through the North Dunes or along the beach towards Horsey. Further up the coast there is another beach car park, managed by the National Trust, at Horsey Corner. To access the car park turn off the B1159 at Horsey Corner. From here walk south along the public footpath on the inland side of the dunes or, better still, along the beach. A further, alternative point of access is along the public footpath from the Nelson Head pub in Horsey village.

Site description

Grey and Common Seal: Anywhere along this stretch of coast can produce sightings of both species but probably the best time to observe them is during breeding, which takes place from late summer through the autumn for Common Seals, and from October through December for Grey Seals. During these times please stay well away from the animals as unnecessary disturbance is likely to affect their behaviour and breeding performance. The largest gatherings are usually found on the beach between Winterton Ness and where the public footpath from the Nelson Head pub reaches the beach.

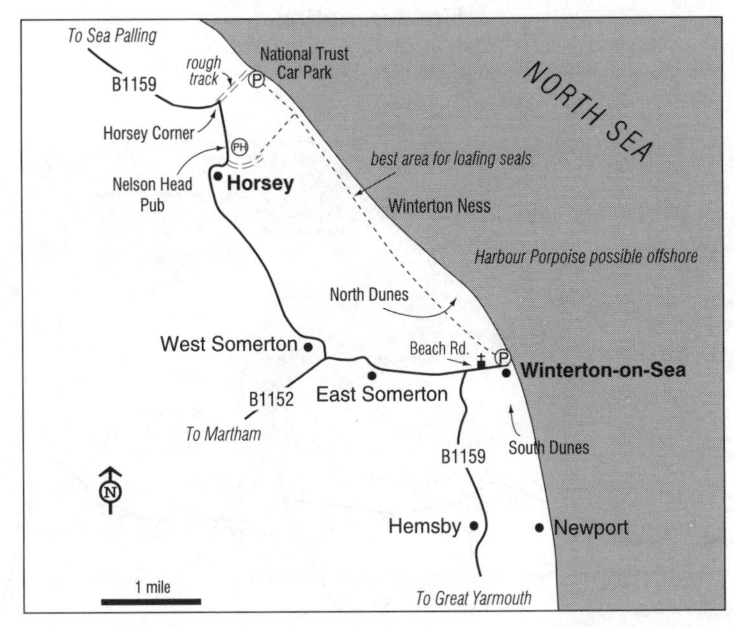

Harbour Porpoise: Harbour Porpoises are most commonly seen off Winterton-on-Sea, although they may be observed from any point of the coast.

Stoats are common in the dunes north of Winterton.

Other information

In the last few summers the Little Tern colony on Winterton-on-Sea beach has increased due to displacement from the Great Yarmouth colony. Each year, staff from the Royal Society for the Protection of Birds erect fenced enclosures to protect the birds from disturbance, and it is important that these restrictions are observed.

16 Blakeney Point and Cley NWT Reserve, Norfolk

Grid ref: TG005465
Grid ref: TG053444

Key species

Common Seal, Grey Seal, Brown Hare, Water Vole, Chinese Water Deer.

Access

Blakeney Point can be accessed on foot from Cley beach car park to the east: a walk of about four miles (6km), or by boat from Morston or Blakeney. The Point is open throughout the year but access is restricted to certain areas when a large colony of terns is nesting between April and September. Boats to Blakeney Point leave from Morston (TG009438) and from Blakeney Quay (TG027440). Visit www.glavenvalley.co.uk/ for further details. The seal colony can also be viewed, albeit distantly, from the footpath between Morston and Stiffkey.

Site description

The National Nature Reserve of Blakeney Point is a long sand and shingle spit. Research suggests that the Point is moving landwards at a rate of a metre per year.

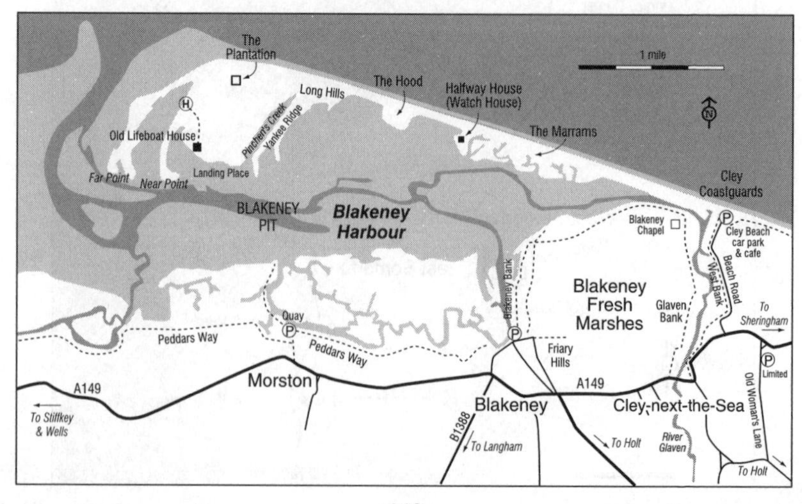

Common Seal and Grey Seal: The North Norfolk coast, west to the Wash, holds the largest breeding colony of Common Seals in the UK, representing about 7% of the total population. The best place to observe large numbers of animals is at Blakeney Point. The colony resides just off the Point upon the extensive sand flats. About 400 strong, the colony mainly consists of Common Seals, with smaller numbers of Grey Seals. Although it is possible to observe good numbers of seals at any time of year, the best times are during breeding and pupping. Common Seals pup between June and August. Grey Seals are later: they generally give birth between October and December.

Brown Hare: Brown Hares are common on the landward side of Blakeney Point and can also be seen in the Eye Field by Cley Beach car park.

The extensive reedbeds, scrapes and wet grassland that make up Cley are home to a number of mammal species. Permits to visit the reserve can be obtained from the NWT visitor centre off the A149 coast road. **Water Voles** are relatively easy to see in the ditch by the A149 in front of the visitors' centre. Small numbers of **Chinese Water Deer** have recently colonised the area. They are most frequently seen from Dauke's Hide late in the day. **Muntjac** are also regular and **Otters** are occasionally seen. Nearby, Stoats are regularly seen at Salthouse village, particularly around Gramborough Hill (TG082439).

Other information

The vegetation and bird populations of Blakeney Point are extremely vulnerable to disturbance. Visitors are asked to keep to well-worn paths and to keep away from the tern breeding colonies that are fenced off between April and September. An information centre at Morston Quay provides further details on the area's attractions. For more information see http://www.cley.org.uk/to_the_point.htm

17 Holkham, Norfolk

Grid ref: TF891440

Key species

Muntjac, Fallow Deer, Brown Hare.

Access

Holkham is on the A149 approximately two miles (3km) west of Wells-next-the-Sea on the north Norfolk coast. Lady Ann's Drive runs north to south from the conifer belt through the village to Holkham Park. Car parking along Lady Ann's Drive costs £2.50 during the summer and at weekends at all other times. Alternative free parking may be found south of the A149 in Holkham village and at nearby Burnham Overy. The whole area is criss-crossed by public footpaths.

Site description

Holkham National Nature Reserve covers 4,000 hectares with the core area lying between Wells-next-the-Sea and Holkham Bay. During the nineteenth century pine trees were planted on the dunes to create a shelter belt, protecting the reclaimed farmland from wind-blown

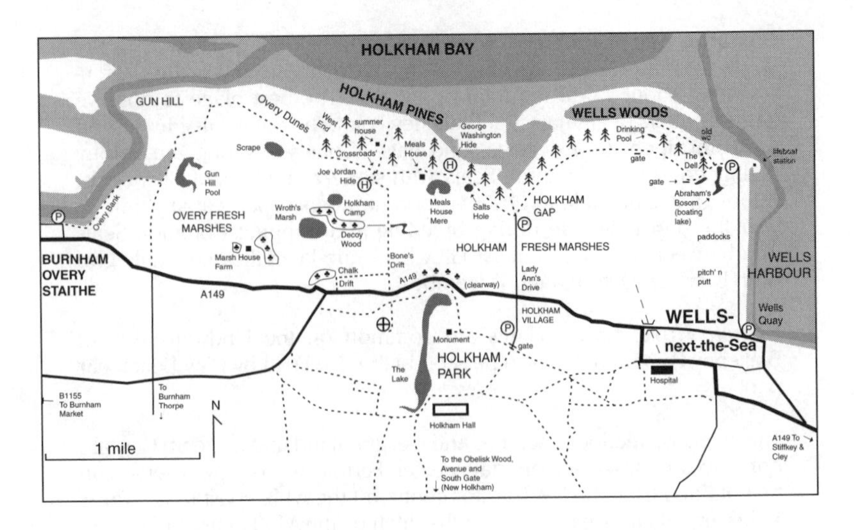

sand. The diverse range of habitats that the reserve encompasses also includes sand dunes, pasture and arable land, and both fresh and salt marshes.

Muntjac: The pine belt at Holkham supports a large population of Muntjac, and they are easily seen throughout the day, but the best time is early in the morning when there is less disturbance. The section west of Lady Ann's Drive is the most productive, particularly around Meals House (TF880451): the animals often come out onto the path to graze and browse the trackside vegetation.

Brown Hare: Brown Hares are common on Holkham Estate and may be observed on the fresh marsh from either the Washington or Joe Jordan Hides or on the adjacent arable farmland.

Fallow Deer: Holkham Park is a wonderful location in which to view parkland Fallow Deer. Approximately 600 deer currently reside within the 1,200-hectare park. Access is south along Lady Ann's Drive in Holkham village.

Other species

As may be expected on such a large and diverse estate, Holkham supports sizeable populations of many common mammals. **Water Voles** occur on the fresh marsh, although viewing them is extremely difficult due to distance from the watch points and hides. Common mustelids such as **Weasels** and **Stoats** can often be observed from the hides, particularly during the bird-breeding season when they have a tendency to hassle ground-nesting species such as Lapwings. **Grey Squirrels** are common within the pine belt. **Daubenton's Bat** is common over the lake in Holkham Park.

Other information

During winter the fresh marsh at Holkham supports a massive pre-roost of Pink-footed Geese. To witness the many thousands of birds flying in off nearby farmland is to observe one of Britain's great wildlife spectacles.

MIDLANDS

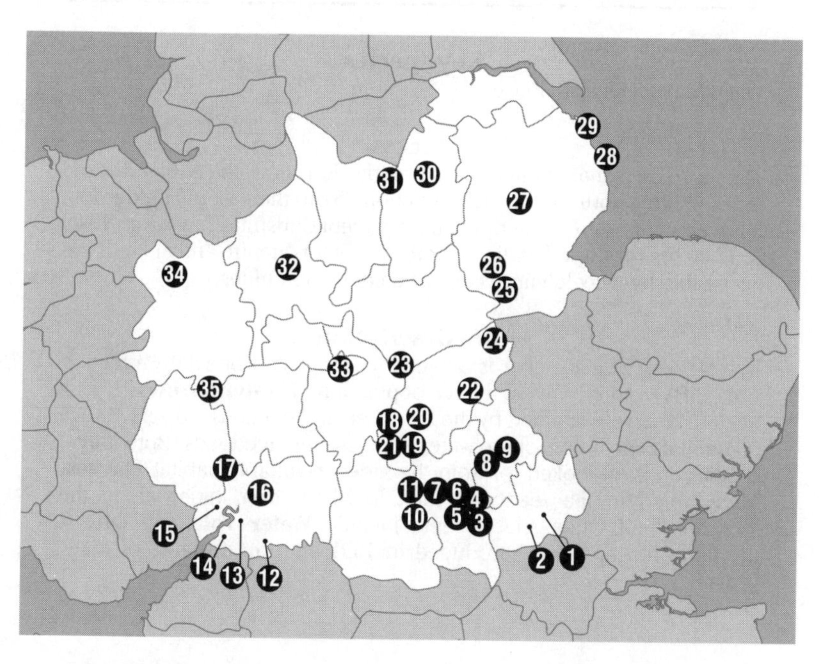

1 Rye Meads RSPB Reserve
2 Broxbourne Woods NNR
3 Ashridge Park
4 Wendover Woods
5 Tring Reservoirs
6 Stockgrove Country Park
7 Thornborough Bridge
8 Woburn Abbey and Park
9 Ampthill Park
10 Boarstall Duck Decoy
11 Whitecross Green Wood
12 Cotswold Water Park
13 Woodchester Park
14 Slimbridge WWT
15 Wye Valley and Forest of Dean
16 Over Hospital
17 Dymock Woods
18 Everdon Church
19 Bucknell's Wood
20 Grand Union Canal
21 Fawsley Lakes
22 Barnwell Country Park
23 Wardley Wood
24 Rockingham Forest
25 Bourne Woods
26 Callan's Lane Wood
27 Hartsholme Country Park and Swanholme Lakes SSSI
28 Saltfleetby-Theddlethorpe Dunes NNR
29 Donna Nook NNR
30 Sherwood Forest NNR
31 Elvaston Castle Country Park
32 Cannock Chase
33 Cannon Hill Park, Moseley
34 Montfod Bridge
35 Wyre Forest

1 Rye Meads RSPB Reserve, Hertfordshire

Grid ref: TL383101

Key species

Water Shrew, Harvest Mouse.

Access

The reserve is situated three miles (nearly 5km) from the A10 and seven miles (11km) north of the M25 (junction 25), to the east of Hoddesdon. It is well signposted and is open daily (except Christmas Day) from 10:00 to 17:00 hrs (or dusk if earlier). There is a visitors' centre and all trails are accessible by wheelchairs as are seven of the ten hides.

Site description

Rye Meads RSPB Reserve is a recently designated Special Protection Area (SPA) and is home to a large population of **Water Shrews**. This 40-hectare reserve is situated by the River Lee and contains a diverse range of habitats including open water, scrapes and reedbeds. Both target species are best looked for from the hides in suitable habitat. **Harvest Mice** breed in the reedbeds; look for them from hides along the Moorhen Trail. Other species here include **Water Vole** and **Otters**, the latter having been introduced in 1991 after an absence of many years.

Harvest Mouse (*P. Twisk*)

2 Broxbourne Woods NNR, Hertfordshire

Grid ref: TL324071

Key species

Muntjac, Weasel.

Access

The woods are 1.5 miles (more than 2km) from Brickendon village and there are car parks at Broxbourne and Bencroft woods. Picnic facilities are available within the car parks. Various tracks dissect the woods, some of which are passable to wheelchair users.

Site description

These woods, managed by the Woodland Trust and Hertfordshire County Council in conjunction with Natural England, form part of the largest wooded area in Hertfordshire, covering a considerable land area in the south-east of the county. Hornbeams are common and the Sessile Oaks are of an eastern variant. **Muntjac** and **Grey Squirrels** are commonly seen, particularly at dawn and dusk. Other species that occur here include **Weasels** and **Badgers**.

3 Ashridge Park, Hertfordshire/ Buckinghamshire

Grid ref: SP970130

Key species

Edible Dormouse, Fallow Deer.

Access

The area is easily accessed via minor roads north-west of Hemel Hempstead. There is a car park at the above grid reference.

Site description

This large estate runs along the main ridge of the Chiltern Hills. There are endless forest trails and a visitors' centre with parking to allow you to get into the heart of this spectacular area in search of the **Edible Dormice** that are present throughout in suitable habitat. **Fallow Deer** and **Muntjac** are common throughout.

4 Wendover Woods, Buckinghamshire

Grid ref: SP886105

Key species

Edible Dormouse.

Access

The woodlands are located along the B4009 between Wendover and Tring. Take the right-hand turn approximately one mile (1.6km) north of RAF Halton, signposted Wendover Woods and St Leonard's. The main entrance to the woods is on the right near the top of the hill opposite the golf club entrance. There is a car park with picnic area and toilets. The gates are closed at dusk but the woods can still be accessed on foot after dark. Alternatively take the first right immediately after RAF Halton and park after about 200 yards at the barrier across the road. Walk past the barrier and along the path until you come to a clearing on your left with Yew trees to your right. You should see trees damaged by Edible Dormice all around you.

Site description

This is a Forestry Commission woodland situated on the northern edge of the Chiltern escarpment. It is a large mixed woodland of some 325 hectares. **Edible Dormice** are common within the woodlands and can be heard and occasionally seen at the start of the main trail. However, to maximise your chances, walk the first 300 yards of the main trail and any of the side trails with a torch – being careful not to spotlight too close to any buildings given the proximity of the RAF camp! It is not unusual to hear ten or more animals in a fairly small area but you normally have to be quick to see them. They tend to avoid the spotlight although some success may be had if you use a red filter. **Badgers** are also common in the same area, whilst the woodlands also support **Muntjac,** which are often seen browsing vegetation along the main trail, as well as **Red Foxes** and **Weasels**.

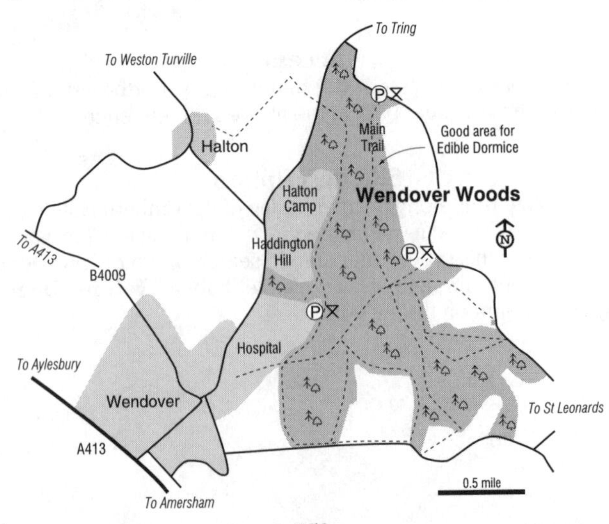

5 Tring Reservoirs, Buckinghamshire/ Hertfordshire

Grid ref: SP918140

Key species

Soprano Pipistrelle, Common Pipistrelle, Noctule, Serotine, Daubenton's Bat.

Access

From the A41 (T) Tring to Aylesbury road, follow the B489 towards Marsworth. Go past Wilstone Reservoir to a roundabout; go straight ahead and on entering the village of Marsworth the entrance to the reservoirs is on the right. There is a pay-and-display car park with information boards. A footpath leads to the Grand Union Canal and reservoirs.

Site description

The site comprises three reservoirs surrounded by coniferous woodland and pasture. A footpath leads between two of the reservoirs and provides a good vantage point for observing bats.

Other information

The North Bucks Bat Group and Friends of Tring Reservoir often organise a bat walk each year during European Bat Weekend (August Bank Holiday weekend). Visit the Friends of Tring Reservoir website (www.fotr.org.uk) for further information.

6 Stockgrove Country Park, Bedfordshire/ Buckinghamshire

Grid ref: SP916291

Key species

Soprano Pipistrelle, Common Pipistrelle, Noctule, Daubenton's Bat, Brown Long-eared Bat.

Access

The park entrance is on the minor road between Heath and Reach and Great Brickhill (SP920294). At the roundabout on the A5 between Hockliffe and Little Brickhill, take the minor road signed Heath and Reach, and then take the first road on the right. The park is signposted. There is a car park but this is locked during the evening. However, there is room to park on the roadside verge just outside the park. A path from the car park follows the valley and leads to the lake.

Site description

The park comprises a mix of habitats, with deciduous and coniferous woodland, acid grassland and wet areas. A lake provides the focus for the bat interest. Part of the woodland (Bakers Wood) is an SSSI, and King's Wood NNR is immediately across the road to the north. The park, which is managed by the Greensand Trust for the Bedfordshire and Buckinghamshire County Councils, is fully accessible to the public, as is the adjacent Oak Wood.

Both **Common Pipistrelles** and **Soprano Pipistrelles** are abundant around the lake and in the surrounding woodland. **Noctules** roost in trees in and around the country park, and there are frequently two or more feeding high above the lake shortly after sunset. **Daubenton's Bats** roost in the woodland surrounding the park and there are usually a number of individuals feeding low over the water about an hour after sunset. Other bat species that have been recorded in the area include **Brandt's Bats**, **Natterer's Bats, Nathusius's Pipistrelles** and **Barbastelles**. Other mammal species present include **Badgers**, **Muntjacs**, **Grey Squirrels**, **Water Shrews** and **Brown Rats**.

7 Thornborough Bridge, Buckinghamshire

Grid ref: SP730332

Key species

Common Pipistrelle, Soprano Pipistrelle, Daubenton's Bat.

Access

Thornborough Bridge is about 1.5 miles (more than 2km) east of Buckingham on the A421 Buckingham to Milton Keynes road. It is well signposted with a small car park.

Site description

This Roman bridge is surrounded by pasture on all sides and a public footpath runs alongside the river for nearly a mile to the Coombs quarry. **Pipistrelles** and **Daubenton's Bats** regularly feed around the bridge and along the water to either side. **Noctules** can sometimes be seen over the pastures.

8 Woburn Abbey and Park, Bedfordshire

Grid ref: SP965327

Key species

Muntjac, Chinese Water Deer, Daubenton's Bat.

Access

The extensive woodlands to the east of Woburn village in Woburn Park have numerous public footpaths that make the area easily accessible. Woburn Park is open daily.

Site description

This is one of the original haunts of both **Muntjac** and **Chinese Water Deer** and both can still be found in the grounds of Woburn Abbey and in the surrounding woodlands. Chinese Water Deer are most easily seen at Whipsnade and in the Woburn area, where they feed out in the fields at dusk. Good areas include the fields adjacent to Steppingly Woods (SP995430) and the woods on the Bedfordshire/Buckinghamshire border on the road from Woburn to Little Brickhill.

The lakes between Woburn and Woburn Park can produce large numbers of **Daubenton's Bats.**

9 Ampthill Park, Bedfordshire Grid ref: TL024383

Key species

Common Pipistrelle, Soprano Pipistrelle, Noctule, Daubenton's Bat, Natterer's Bat.

Access

The entrance to the park is on the B530, close to the junction with the A507 on the western edge of Ampthill. There is a car park at the entrance.

Site description

The park consists of grassland with some scrub and woodland on the scarp of the Greensand Ridge. There are two large ponds below the ridge. The park is owned by Ampthill Town Council and is open to the public. Both **pipistrelle** species are common both on the grassland and in the nearby woodland on the eastern edge of the park. The woodland is also frequented by **Natterer's Bats** and **Noctules**. The large pond north of the car park often has **Daubenton's Bats** flying low over the water about an hour after sunset.

10 Boarstall Duck Decoy, Buckinghamshire Grid ref: SP623151

Key species

Whiskered Bat.

Access

This National Trust property is just off the M40 approximately ten miles (16km) north-east of Oxford and just north of Boarstall village.

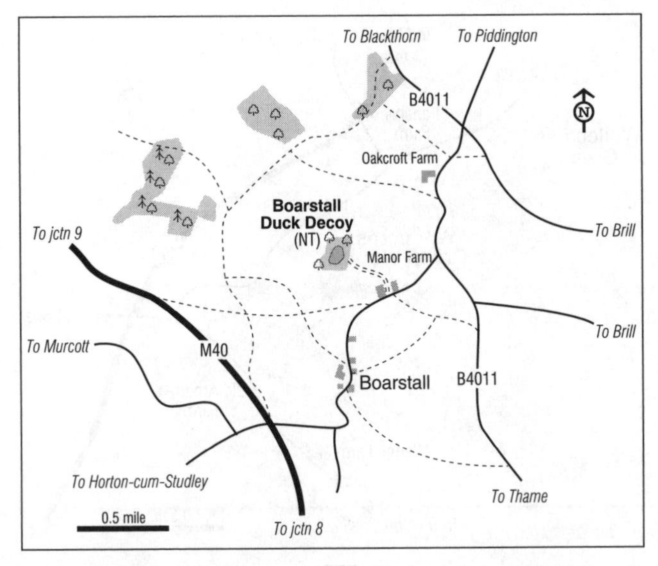

Site description

This tree-fringed lake is a perfect hunting ground for a number of bat species including the **Whiskered Bat**. This species is usually seen darting around the well-wooded edges of the water in search of prey. It is best to consult the National Trust website as the opening hours are very variable. If the property happens to be shut at the time of your visit, there are a number of public rights of way that encircle the pond and should provide success as the bats move around while foraging.

11 Whitecross Green Wood, Oxfordshire

Grid ref: SP602147

Key species

Brown Hare, Muntjac.

Access

This 64-hectare nature reserve is managed by the Berkshire, Buckinghamshire and Oxfordshire Wildlife Trust and is open all year round. To reach the wood from Islip, take the minor road towards Merton, then first right to Murcott. Approximately one mile (1.6km) past Murcott there is a small cottage on the left and the wood is immediately opposite. The woodland rides are generally flat enough for wheelchair access in fine weather, although some of the tracks are quite bumpy.

Site description

Aside from a small section of coniferous plantation, Whitecross Green Wood consists of thickets of blackthorn and hawthorn, as well as hazel coppice and some majestic oak and ash standards. A growing population of **Muntjac** had halted the coppicing operations during the 1990s due to their prodigious browsing of the regrowth, but the Wildlife Trust has started to put into place specialist deer fencing that will allow the age-old prac-

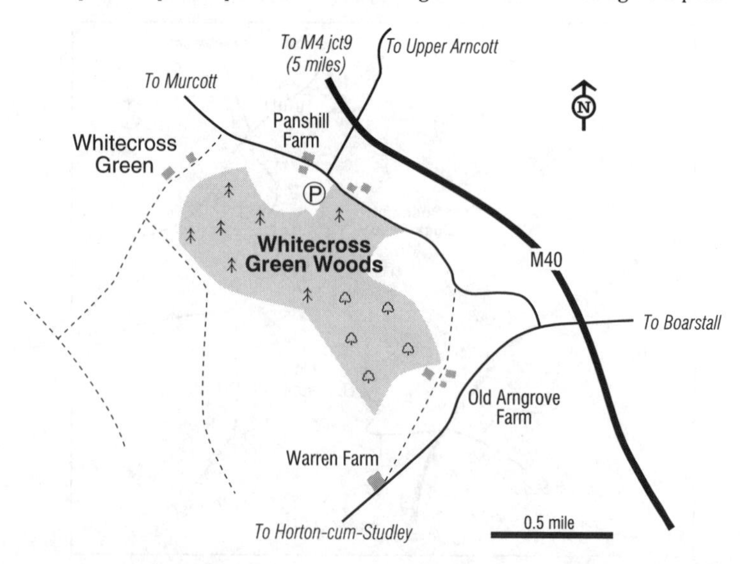

tice to be revived. The wide rides in Whitecross Green Wood are favoured areas for **Brown Hares,** particularly during the mating season in spring.

Other information

This site also supports a number of notable butterfly species including Black Hairstreak, Brown Hairstreak, Grizzled Skipper, Purple Emperor, White Admiral and Wood White.

12 Cotswold Water Park, Gloucestershire/Wiltshire **Grid ref: SU048956**

Key species

American Mink, Otter, Water Vole.

Access

The park is signposted from the A410 between Cirencester and Swindon. There are ample car parking facilities.

Site description

Covering an area of over 100km², with 133 lakes that have been created by gravel extraction, the Cotswold Water Park is Britain's largest water park and is still growing in size. Although much of the area is dedicated to leisure and recreational activities, there are three designated nature reserves within the park that are home to a great deal of wildlife. The complex of pits and lakes that make up the Cotswold Water Park provide perfect habitat for all three species listed.

The following species have also been recorded within the park: **Daubenton's Bat, Common Pipistrelle, Brown Long-eared Bat, Water Shrew, Pygmy Shrew, Common Shrew, Hedgehog, Badger, Brown Hare, Polecat, Stoat, Weasel** and **Roe Deer.**

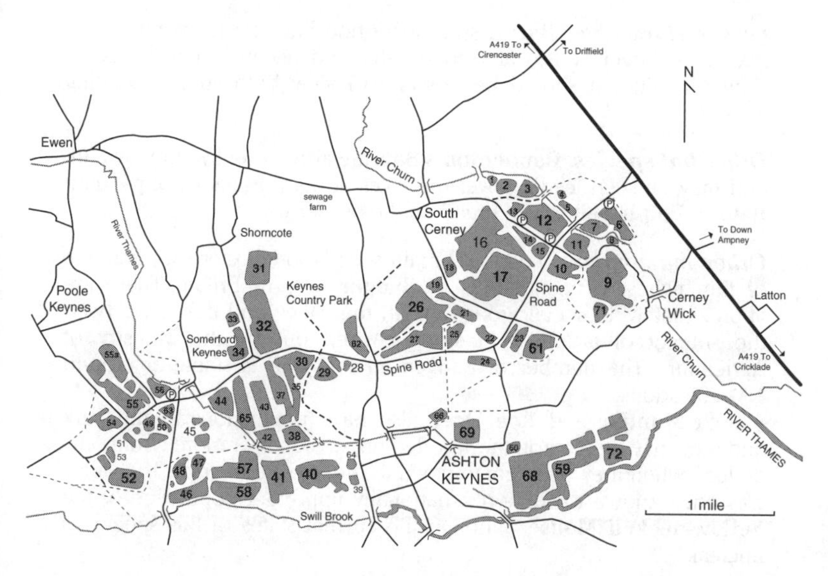

Other information

The nationally rare Snake's Head Fritillary can be found flowering here as well as many orchid species, including Burnt Tip, Early Marsh and Green-winged Orchids. Further information can be obtained at www.waterpark.org

13 Woodchester Park, Gloucestershire

Grid ref: SO822012

Key species

Greater Horseshoe Bat, Lesser Horseshoe Bat, Daubenton's Bat, Badger.

Access

Woodchester Park is signposted off the A46, three miles (nearly 5km) south of Stroud. Access is limited to 09:00–20:00 hrs from May to September and 09:00–17:00 hrs from October to April.

Site description

Greater Horseshoe Bat: Woodchester Mansion and the surrounding parklands are home to the only breeding colony of Greater Horseshoe Bats east of the River Severn and have been home to this species since the early twentieth century. The cattle-grazed grasslands within the park provide ideal foraging habitat for this species. Although numbers have varied greatly over the past few decades, the introduction of cattle to the park has recently seen an increase to approximately 150 individuals in 2003.

If the park is closed at dusk public footpaths through Stanley Wood to the north and in and around High Wood to the south may provide an opportunity to observe foraging Greater Horseshoe Bats as they disperse to feed.

Lesser Horseshoe Bat: Lesser Horseshoe Bats are more numerous than their larger cousin at Woodchester, and because they forage in similar habitats it should be possible to view both species feeding together.

Other bat species: **Daubenton's Bats** are also present within the park and may be seen feeding over the lakes at dusk. **Brown Long-eared Bats** and **pipistrelles** also breed on the property.

Other mammals: Home to 12 main setts, Woodchester Park has one of the highest concentrations of **Badger** setts in Britain. There are approximately 15 social groups within the valley and they have been the subject of a long-term study carried out by the Ministry of Agriculture. The number of Badgers in the park is stable at around 150–200 adults.

Both **Muntjac** and **Roe Deer** also have populations in the park and they may be seen on the wooded upper slopes or more commonly at dusk, when they often come down to the pasture to graze. The valley was once known to support a nationally important population of the **Yellow-necked Mouse**, although the current status of this species is unclear.

14 Slimbridge WWT, Gloucestershire

Grid ref: SO722048

Key species

Water Vole.

Access

Slimbridge is between Bristol and Gloucester and is signposted off the M5 at junctions 13 and 14. The reserve is open daily (except Christmas Day) from 09:30 to 17:00 hrs (16:00 hrs from November to March). Admission charges for non-members are: adults £6.75, children £4, senior citizens £5.50.

Site description

Opened in 1946, this world famous wildfowl and wetland reserve has developed some exceptional visitor facilities making for a rewarding day out. **Water Voles** are common and can easily be viewed from many of the hides. Dykes and ponds where the bankside vegetation is thick usually offer the best opportunities for sightings.

Other information

This reserve is most famous for its fantastic gatherings of wildfowl during the winter when it hosts large numbers of wild geese, swans and ducks.

Water Vole (*M. Nāzāreanu*)

15 Wye Valley and Forest of Dean, Gloucestershire/Monmouthshire

Grid ref: SO655141

Key species

Greater Horseshoe Bat, Lesser Horseshoe Bat, Fallow Deer.

Access

The Forest of Dean is well signposted from all directions and good visitor facilities make this site very accessible. Various sites within the forest are good for mammals and access details for each are given below.

Site description

Traditionally an oak forest, the Forest of Dean now consists of an even mix of broadleaved and coniferous tree species, although the main conservation interest lies within the 1,000 hectares of ancient oak woodland remaining in the Cannop Valley. There are currently ten SSSIs within the forest that are managed by the Forestry Commission, and a further 40 local nature reserves managed in partnership with the County Wildlife Trusts.

Greater Horseshoe Bat: This complex of sites on the border between England and Wales represents the northern edge of the range of the Greater Horseshoe Bat. The site contains the main maternity site for the species in this area, about 6% of the total UK population. The bats are believed to hibernate in disused mines within the forest.

Lesser Horseshoe Bat: The Forest of Dean and Wye Valley woodlands contain by far the greatest concentration of Lesser Horseshoe Bats in the UK, in total about 26% of the national population. The majority of sites within the area are maternity roosts. Like the previous species, the Lesser Horseshoe Bat is believed to hibernate in the many disused mines in the area.

Lady Park Wood

Both Greater and Lesser Horseshoe Bats are present within Lady Park Wood NNR. Indeed, the importance of this woodland cannot be underestimated, as the former species is present in nationally important numbers, with the latter here in internationally important numbers. Lady Park Wood has been recognised as one of the most important woodland conservation areas in Great Britain, and permits (obtained from Natural England) are required for access.

Fallow Deer: Fallow Deer are quite common within the Forest of Dean, where a population of around 400 individuals can be regularly seen. Wildlife rangers organise deer safaris in February and October each year that provide an opportunity to learn more about the deer. For more details please refer to the Forestry Commission website (Forest of Dean section: www.forestry.gov.uk). Fallow Deer may also be seen in the woodlands at Symonds Yat Rock (SO563160), near Coleford, and at Nagshead RSPB Reserve (SO607085).

Noctule and pipistrelle species: These species are quite common in the Forest of Dean, and a good place to view them is at Speech House woodland near Cinderford (SO624125). This SSSI consists of an area of ancient trees that provides ideal conditions for roosting bats and for the many species of uncommon lichen and moss that occur. The woodlands are located to the north of the B4226 Cinderford to Coleford road, approximately 500m from Speech House.

Badger: Badgers are common within the Forest of Dean, and a good place to find them is the Nagshead RSPB Reserve (SO606085). Access to the reserve is from Coleford from the B4431 heading towards Parkend. The reserve is signposted to the left just before Parkend village and is accessed along a forest road. There is a small car park, and a daily bulletin board lets you know which bird species may be seen at the site. Badgers also occur in the woodlands around Symonds Yat Rock, which is well signposted off the B4228 (SO563160).

Other information

Other species that occur in the area include the **Otter**, on nearby stretches of the River Wye, **Brandt's Bat**, in woodlands just south-east of Drybrook, **Brown Long-eared Bat**, **Daubenton's Bat, Whiskered Bat, Natterer's Bat, Hazel Dormouse, Wood Mouse, Yellow-necked Mouse, Field Vole, Bank Vole, Common Shrew, Pygmy Shrew, Water Shrew, Mole, Hedgehog, Red Fox, Stoat, Weasel, American Mink** and **Muntjac**. Since November 2004, groups of up to ten **Wild Boars** have also been seen in the Forest of Dean, including a sow with piglets spotted in a private garden. The source of these animals remains unknown, although an unofficial deliberate release seems likely.

16 Over Hospital, Gloucestershire

Grid reference: SO807195

Key species

Lesser Horseshoe Bat.

Access

The now disused Over Hospital lies north of Gloucester and is signposted off the A40. Park in front of the hospital and knock at the main hospital doors to get permission to enter.

Site description

Security guards are present but they are normally friendly. Walk behind the main hospital building and amongst a series of small buildings/garages you will see a new boiler room that has been built to replace the previous dilapidated building. The new building has been designed with bats in mind, and the colony of Lesser Horseshoe Bats has successfully completed the switch. The maximum count since the new building was created is 43. A bat detector is recommended as once the bats emerge they quickly seem to disappear into the hedgerow. **Noctules** and **Brown Long-eared Bats** are often seen flying high over the buildings at this site.

17 Dymock Woods, Gloucestershire

Grid ref: SO6772885

Key species

Hazel Dormouse, Fallow Deer.

Access

Leave the M50 at junction 3, take the B4221 towards Kilcot and then follow signs to Kempley. There is a car park and several trails.

Site description

This is a mixed woodland with many outstanding Sessile Oaks. The **Hazel Dormouse** is present throughout with good numbers in areas of suitable habitat. **Fallow Deer** may also be seen here, and the woods also support a small population of **Muntjac**.

Other information

A visit to this site in early spring will provide an exceptional display of wild daffodils. White Admirals and Nightingales also occur here.

18 Everdon Church, Northamptonshire

Grid ref: SP594574

Key species

Natterer's Bat.

Access

Everdon is approximately five miles (8km) south of Daventry.

Site description

The **Natterer's Bat** roost at Everdon Church was discovered in 1981 in the disused south porch. The bats enter and leave the roost via a hole between the wooden roof beams and the church wall. Over 60 bats have been present each year since 1998 and this is the largest and most important roost of this species in Northamptonshire.

Natterer's Bat (*D. Roth*)

19 Bucknell's Wood, Northamptonshire

Grid ref: SP650440

Key species
Brown Long-eared Bat, Noctule, Common Pipistrelle, Soprano Pipistrelle.

Access
There is a small Forestry Commission car parking area at the south-eastern edge of the wood. From the A43 turn into and go through Silverstone village and continue for a mile (1.6km) to the car park.

Site description
This semi-natural ancient woodland supports many mammal species. It is managed by the Forestry Commission who are in the process of thinning out the conifers to restore the wood to its former glory. All the target species can be seen throughout the wood in summer.

20 Grand Union Canal, Northamptonshire

Grid ref: SP741500 (Stoke Bruerne)

Key species
Daubenton's Bat.

Access
Numerous roads cross over the canal and provide opportunities to view feeding bats. The Grand Union Canal Walk runs adjacent to the water for a considerable distance.

Site description
Running between Northampton and Milton Keynes, this section of canal supports good numbers of **Daubenton's Bats**. The best access points to the Grand Union Canal are at Stoke Bruerne, Grafton Regis and Bugbrooke. Daubenton's Bats are regularly seen feeding along the canal at these points.

21 Fawsley Lakes Northamptonshire,

Grid ref: SP565569

Key species
Common Pipistrelle, Soprano Pipistrelle, Daubenton's Bat, Whiskered Bat.

Access
Fawsley Lakes are three miles (nearly 5km) south of Daventry. Head east along a single-track road off the A361 about a mile (1.6km) south of Badby. After 1.5 miles (2.5km) pass Fawsley Hall on the right and park on the right near the gate entrance to the church. Walk back along the road to the roadside lake.

Site description

Ornamental lakes set in an ancient parkland of a stately home. The grounds and parkland are private but criss-crossed with footpaths. **Noctules** are occasionally seen, especially in spring and early summer. **Brown Long-eared Bats** roost in the hall and are occasionally seen foraging in lakeside vegetation. **Nathusius's Pipistrelle** has been recorded once.

22 Barnwell Country Park, Northamptonshire

Grid ref: TL033872

Key species

Whiskered Bat.

Access

The park is on the southern outskirts of Oundle. Admission is free, although there is a small charge for use of the car park. The park is open 24 hours a day, all year. The visitors' centre is open 11:00–17:00 hrs at weekends and during school holidays.

Site description

This park offers waterside walks, nature trails and bird hides. Situated in the Nene Valley water meadows and consisting of a complex of lakes, pools and meadows its 37 hectares provide an opportunity for viewing **Whiskered Bats**. During summer, they can be seen hunting along the tree-lined edge just north of the car park. Other bat species present include **Soprano Pipistrelle**, **Common Pipistrelle**, **Daubenton's Bat** and **Noctule**.

Other mammals include **Water Shrews** and the occasional **Otter**.

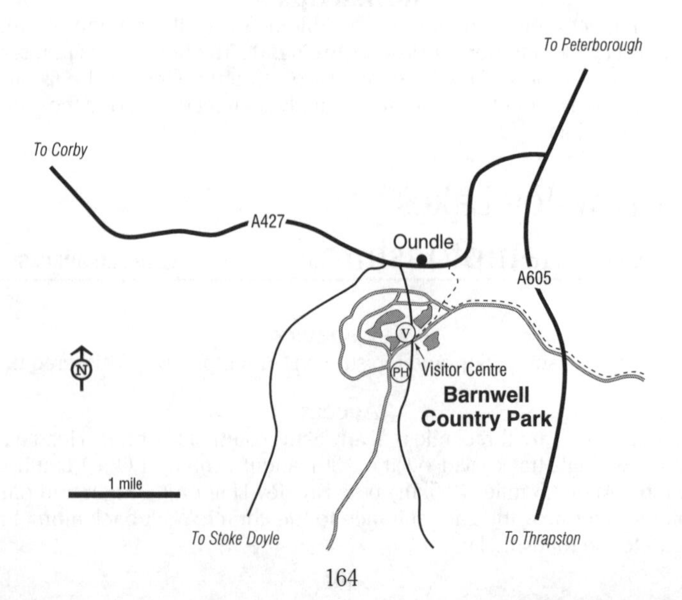

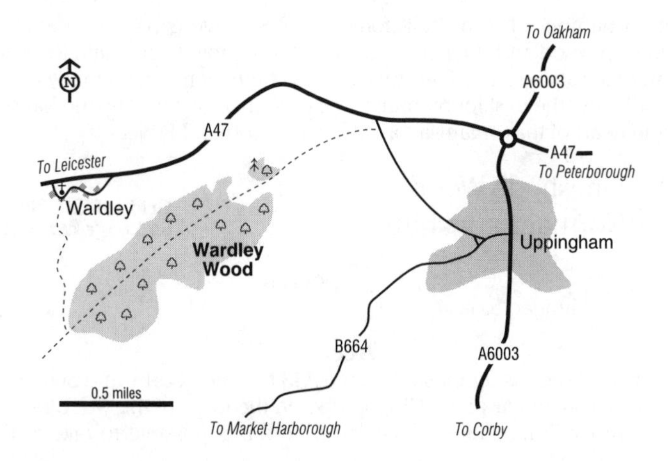

23 Wardley Wood, Leicestershire

Grid ref: SP833992

Key species

Muntjac, Badger, Red Fox.

Access

Wardley Wood can be accessed via public footpaths from either Uppingham village or Wardley village.

Site description

This is an ancient semi-natural woodland with areas of conifer plantations, managed by the Forestry Commission. The wood is host to a large population of **Badgers** as well as small numbers of **Red Foxes**. **Muntjac** also occur here. Follow the rides early in the morning and you may see one trotting away in front of you.

24 Rockingham Forest, Northamptonshire

Grid ref: SP863918 (Rockingham Village)

Access

Over 500 miles (805km) of footpaths and bridleways, described in forest guides, provide access to the forest.

Site description

Rockingham Forest covers an area of some 500km^2, bounded by the Rivers Well and and Nene to the east and west, and the towns of Stamford and Kettering to the north and south respectively. The Royal Forest of Rockingham was once a medieval hunting ground, designated as such by William the Conqueror some 900 years ago, and it has always consisted of a diverse patchwork of ancient woodlands and open agricultural land with a number of stone-built forest villages. Large expanses of ancient ridge-top woodland still exist today and these provide the main focus for mammal watchers. The Forestry Commission now manages many of the woodlands (the best for mammals are described below) and the forest is at the heart of that organisation's Ancient Woodland Project.

24a Fineshade Wood, Northamptonshire

Grid ref: SP980983 (Top Lodge car park)

Key species

Roe Deer, Muntjac, Stoat.

Access

Fineshade Wood is signposted off the A43 between Corby and Stamford. There is a small car park at Top Lodge on the edge of the wood beside the Forestry Commission office. From here it is possible to take a trail through the wood.

Site description

This semi-natural ancient woodland made up of many species of native broadleaves is home to numerous mammals. Roe Deer may be seen feeding quietly among the oak and ash trees or in the section of coniferous woodland. Muntjac and Stoats are common.

24b Fermyn Woods Country Park, Northamptonshire

Grid ref: SP955850

Key species

Natterer's Bat, Whiskered Bat.

Access

This area of the Rockingham Forest is signposted from the A6116 at Brigstock. The visitors' centre is open most days and several trails run through the country park and woodlands.

Site description

This country park consists of ancient woodland with semi-natural oak and ash woods, along with conifer plantations. A number of bat species, including Natterer's Bats, feed within the ancient semi-natural woodland in the country park. Whiskered Bats may be observed foraging within and over the woodlands in the park. Many trails may be accessed from the pay-and-display car park.

Whiskered Bat (*Dan Brown*)

25 Bourne Woods, Lincolnshire

Grid ref: TF076201

Key species

Natterer's Bat, Whiskered Bat, Leisler's Bat, Fallow Deer.

Access

These ancient woodlands can be accessed either by public footpath, from Beech Avenue, Bourne; or by road: a car park is signposted off the A151 half a mile west of Bourne. The Forestry Commission has created a series of way-marked trails and well-surfaced paths that radiate out from the car park. The woodland attracts around 100,000 visitors each year.

Site description

The woodlands at Bourne were originally part of the primeval Brunswald Forest, and there has been continuous woodland at the site for at least 8,000 years. The wood now covers some 160 hectares and is owned by the Forestry Commission. The present mix of broadleaved trees and coniferous species is managed for conservation as well as recreation and timber production. Bourne Woods are a productive hunting ground for many species of bat including difficult-to-see species such as **Natterer's Bat, Whiskered Bat** and **Leisler's Bat**. All these are best observed flying over the woodland clearings, along wide rides or over the pools. **Brown Long-eared Bats** and **pipistrelles** also occur at this site.

In 1972 two small lakes were created by damming a small stream deep in the woods. They have come to form a favoured watering hole for the large population of **Fallow Deer** within the wood. These animals' drinking habits are most predictable in the summer when they come down to the water at dawn and dusk each day. The pools are also a good place to look for hunting **Daubenton's Bats**. Other species that may be encountered include **Muntjac, Red Foxes, Badgers** and the elusive **Hazel Dormice**.

26 Callan's Lane Wood, Lincolnshire

Grid ref: TF060270

Key species

Muntjac, Weasel, Stoat.

Access

This small ancient woodland is located one mile (1.6km) west of Kirkby Underwood and five miles (8km) north of Bourne, just off the A15. There is a car park just off the minor road. A hard-surface track runs through the centre of the wood, allowing year-round access to both walkers and cyclists.

Site description

This relatively small woodland, currently a mix of broadleaved and coniferous trees and managed by the Forestry Commission, is gradually being restored under the 'Ancient Woodland Project'. **Muntjac** are present in small numbers and are often seen browsing along the track edge at dawn. **Weasels**, **Stoats** and **Red Foxes** are the main mammalian predators in the wood preying upon the large numbers of **Rabbits** in the area.

27 Hartsholme Country Park and Swanholme Lakes SSSI, Lincolnshire

Grid ref: SK948697

Key species

Common Pipistrelle, Soprano Pipistrelle, Daubenton's Bat.

Access

The main entrance to the park complex is from Skellingthorpe Road and is well signposted. From the car park follow the path in front of the old stable block that contains the warden's office into the wood-pasture area. Follow the path through the trees for a short distance to the boathouse and footbridge. The park can also be accessed from numerous footpaths from the surrounding areas of housing.

Site description

This site comprises a mixture of habitats including the gardens of Hartsholme Hall (long since demolished), ornamental lakes, lakes created by the sand and gravel extraction industry and wet woodland that has colonised the old sand and gravel workings. Swanholme Lakes SSSI forms the southern part of the site and contains a small remnant of the heathland that was once extensive to the south-west of Lincoln. Both **Common Pipistrelles** and **Soprano Pipistrelles** are common throughout the park and can easily be seen from the boathouse and footbridge over the lake. Numerous roosts are known from the housing estates that surround the park. **Daubenton's Bats** can also be watched from these points.

28 Saltfleetby-Theddlethorpe Dunes NNR, Lincolnshire

Grid ref: TF476912

Key species

Water Vole, Water Shrew.

Access

This National Nature Reserve consists of an extensive area of coastal dunes and foreshore stretching for five miles (8km) between Saltfleet Haven in the north and Mablethorpe North End in the south. The area can be accessed east off the A1031, with six designated car parks providing a good starting point for a visit. These are located at Paradise, Rimac, Seaview, Brickyard Lane, Crook Bank and Churchill Lane.

Site description

Natural England has created various freshwater marsh areas and these support good numbers of **Water Voles** and **Water Shrews**.

29 Donna Nook NNR, Lincolnshire

Grid ref: TF440995

Key species

Grey Seal

Access

Located between Grimsby and Mablethorpe on the Lincolnshire coast, Donna Nook is accessed north of the A1031 from North Somercotes. There is a car park at the beach.

Site description

Donna Nook is one of the most accessible sites in Britain to observe breeding **Grey Seals** during late autumn and early winter. The area is a RAF bombing range, and although each visit is usually accompanied by loud bangs, it has provided protection since the colony became established in the early 1970s; this stretch of coast now supports around 2,000 animals. Donna Nook has one of the best pup survival rates of any colony in the country. Pupping usually starts in November on the edge of the sand dune complex. They are easily observed at this site, usually hauled out on the sand banks or feeding close inshore. The beaches are usually empty by January as the seals spend the majority of their time at sea.

30 Sherwood Forest NNR, Nottinghamshire

Grid ref: SK584532

Key species

Noctule.

Access

There are car and bike parking facilities at the Sherwood Forest Country Park, which is open all year round from dusk to dawn. The park is north of the village of Edinstowe along the B6034. There are numerous footpaths and trails within the forest and around the country park. Parts of the NNR are accessible to less mobile users.

Site description

This nature reserve incorporates 200 hectares of the ancient forest of Birklands, an extensive area of old-pasture woodland and heath on the dry nutrient-poor soils of the Sherwood sandstone. The ancient oaks offer perfect roosting sites for many bat species including **Noctules**. They may be observed hunting back and forth above the forest during twilight.

31 Elvaston Castle Country Park, Derbyshire

Grid ref: SK411332

Key species

Daubenton's Bat.

Access

Elvaston Castle Country Park is just east of Derby off the B5010 Borrowash to Elvaston road. It is also well signposted off the A6 and A52 (M1 junction 25). The park office is open from 09:00 to 17:00 hrs Monday to Sunday, while the grounds are open from 09:00 hrs until dusk all year round. The car park is open in summer from 09:00 to 20:00 hrs and in winter from 09:00 to 17:00 hrs.

Site description

Following the proposal in the Countryside Act of 1968 that 'country parks' should be created to provide improved opportunities 'for the enjoyment of the countryside by the public', Elvaston Castle was soon designated as the first of its kind in Britain in 1970. It is owned by Derbyshire County Council. It spans over 80 hectares of woodland, gardens and open parkland, with a large lake as its centrepiece. Part of the park has been designated as a nature reserve, with a nature trail. The Grade II listed boathouse is a well-known maternity roost site for **Daubenton's Bats**. The bats mainly access the boathouse through the first floor window opening at the northern gable end (lakeside) and either side of the ridge-board at the southern gable-end apex. They have also been seen entering the mooring area through its opening to the lake. Up to 77 individuals have been counted leaving the roost and it is thought that up to 200 might be present in the boathouse.

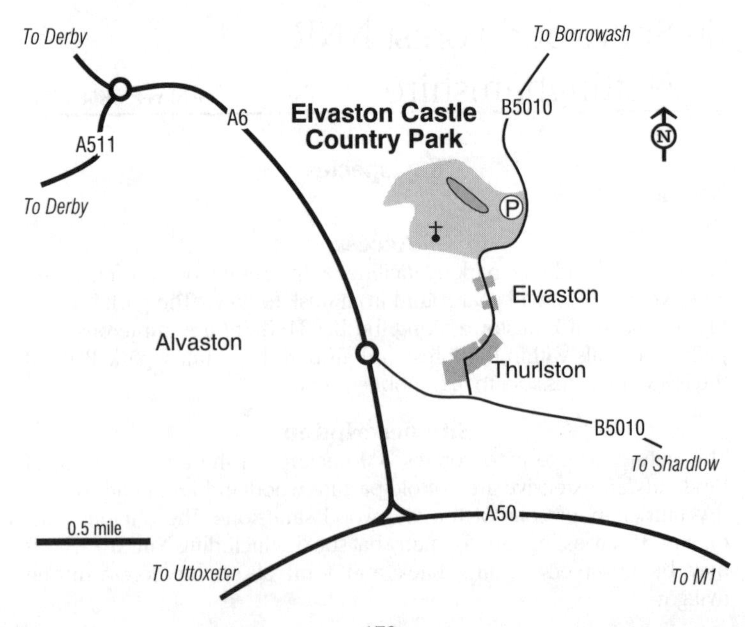

32 Cannock Chase, Staffordshire

Grid ref: SJ990181

Key species

Stoat, Brown Hare, Daubenton's Bat.

Access

Cannock Chase is located between the towns of Rugeley, Cannock and Stafford, and there are many minor roads that bisect the area. There are numerous, well signposted car parks distributed throughout the Chase.

Site description

Once a medieval hunting forest for the Bishops of Lichfield, Cannock Chase is now Britain's smallest designated mainland Area of Outstanding Natural Beauty (AONB), covering 6,880 hectares. Huge numbers of visitors from the massive urban population nearby are placing a great deal of strain on the Chase. The current mix of heathland, planted coniferous woodland and broadleaved woodland has become one of the country's most popular beauty spots and is a huge draw for walkers, cyclists, picnickers and naturalists. Many local schools also use the area for field trips.

The rich variety of habitats that make up Cannock Chase offer the perfect home to many mammal species. **Stoats** are common throughout the area. Look out for this species at the edge of woodlands and along any linear landscape features. **Brown Hares** are uncommon in the Chase, but may be encountered on the open heathland, or bounding across the grassy areas near the young plantations. **Daubenton's Bats** may be found feeding over Stoneybrook Pools (SK018170) during the summer months. Situated in the north-eastern section of Cannock Chase, these woodland pools provide ideal feeding areas for small numbers of this species. It is best to park at the Birches Valley visitors' centre that is easily accessed off the A640 between Hednesford and Rugeley. Other species that may be encountered include **Red Foxes, Badgers, Hedgehogs, Red Deer** and **Fallow Deer**, as well as many smaller mammals.

33 Cannon Hill Park, Moseley, West Midlands

Grid ref: SP060830

Key species

Common Pipistrelle, Soprano Pipistrelle, Noctule, Daubenton's Bat.

Access

The park is situated opposite Edgbaston cricket ground. Park in the car park for the Midland Arts Centre.

Site description

This site consists of parkland dotted with lakes surrounded by areas of woodland. Each of the target species can be seen fairly easily feeding over the parkland and along the woodland edges. The **Daubenton's Bat** population typically forages over the park's water bodies.

34 Montford Bridge, Shropshire

Grid ref: SJ431152

Key species

Noctule, pipistrelles.

Access

Situated three miles (nearly 5km) north-west of Shrewsbury, Montford Bridge village is accessed via the B4380 off the A5. There is a public footpath and bridleway that runs east for half a mile along the south bank of the river.

Site description

The River Severn valley at Montford Bridge is a known foraging area of the **Noctule**. An alternative option to walking the public footpath is to watch from the bridge itself. Either option usually bears fruit with these high-flying bats. **Pipistrelles** also occur at this site.

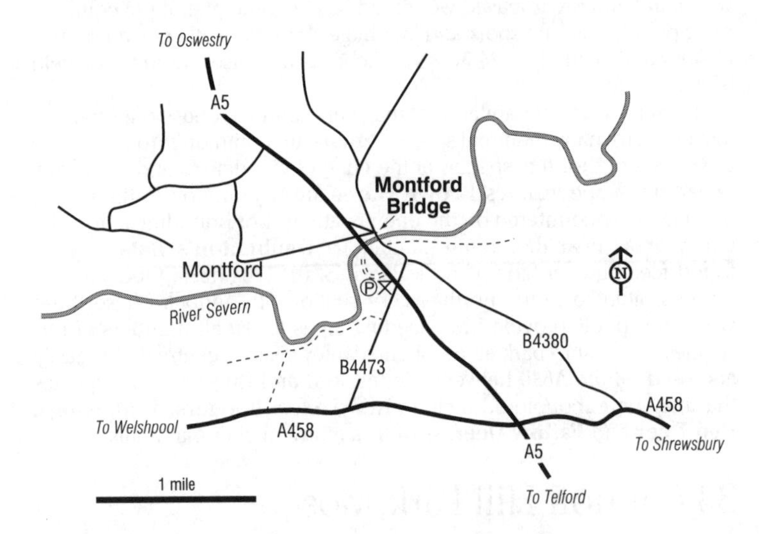

35 Wyre Forest, Worcestershire/ Shropshire

Grid ref: SO750740

Key species

Fallow Deer, Muntjac, Polecat, Daubenton's Bat, Hazel Dormouse, American Mink.

Access

The Wyre Forest Visitor and Discovery Centre is well signposted from all directions. There is also a car park at Earnwood Copse in the north-west corner of the forest on the B4194. There is a large network of public footpaths and bridleways through the forest.

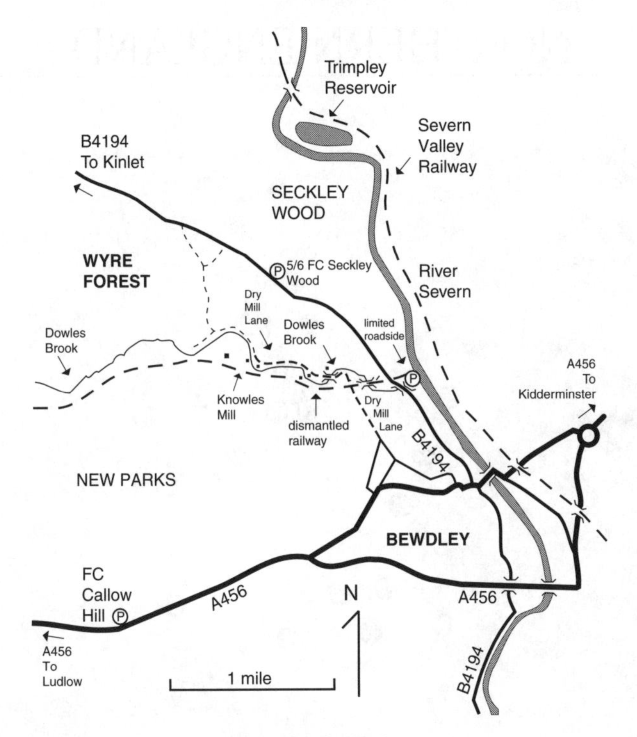

Site description

This beautiful forest, straddling the border of Shropshire and Worcestershire, is one of the three most important areas of ancient woodland in the country. Much of the 2,500 hectares of forest has been designated as a National Nature Reserve and Site of Special Scientific Interest. The present day forest is all that remains of a once extensive tract of woodland that stretched from Worcester up the Severn Valley to Bridgnorth. The Forestry Commission, who manage much of the forest, are currently in the process of reverting many of the non-native coniferous plantations back to broadleaf. They hope to achieve a mix of 85% broadleaf and 15% coniferous woodland.

The forest is an outstanding area for wildlife. Indeed, most British mammals that could be associated with woodland may be found here. **Fallow Deer** are common and are descendants of park Fallow that escaped from Mawley Hall, near Cleobury Mortimer, during the nineteenth century. **Roe Deer** are increasing and may be encountered throughout the woodland. **Muntjac** are also common and increasing, and early morning walks along any of the rides should provide sightings of this animal.

Other mammals in the forest include **Badgers**, **Red Foxes**, **Brown Hares** and **Grey Squirrels**, all of which are common. **Polecats** are currently increasing, whilst **Otters** have started to recolonise the rivers. Dowles Brook is a recommended site for Otters; it may also be possible to find the **American Mink**, particularly where there are areas of grassland and **Rabbits** nearby. **Yellow-necked Mice**, **Hazel Dormice**, **Water Shrews** and various species of **vole** all occur here. Several bats, including both **Common** and **Soprano Pipistrelles** and **Daubenton's Bats**, may be found feeding through the forest during late summer evenings.

NORTHERN ENGLAND

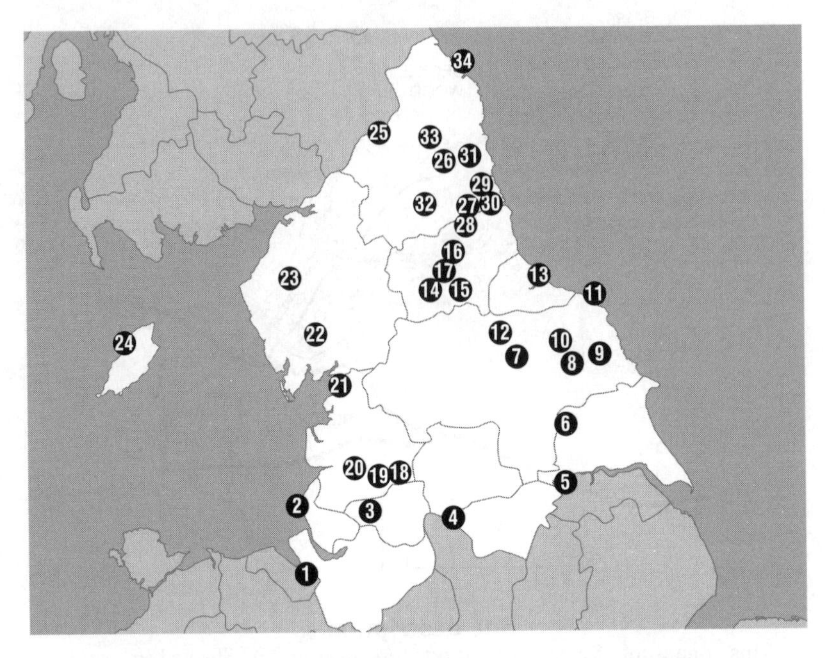

1 Parkgate Marsh
2 Formby Point
3 Pennington Flash Country Park
4 Pennines/Peak District
5 Blackcroft Sands RSPB Reserve
6 Low Catton
7 Boltby Forest
8 St Hilda's Church, Ellerburn
9 Wykeham Forest
10 North Riding Forest Park
11 Upper Derwent
12 Mount Grace Priory
13 Seal Sands
14 Teesdale
15 Raby Castle
16 Malton Nature Reserve and Picnic Site
17 Hamsterley Forest
18 Burrs Country Park, Bury
19 Jumbles Reservoir and Country Park, Bolton
20 Yarrow Valley Park
21 Leighton Moss RSPB Reserve
22 Grizedale Forest Park
23 River Derwent and Bassenthwaite Lake
24 The Curraghs, Ballaugh
25 Kielder Forest
26 Rothley Lakes
27 Shibdon Pond LNR
28 Chopwell Woodland Park
29 Annitsford Pond LNR
30 Whitley Bay
31 Carlisle Park, Morpeth
32 Corbridge
33 Rothbury
34 Farne Islands

1 Parkgate Marsh, Cheshire **Grid ref: SJ282777**

Key species

Water Shrew, Pygmy Shrew.

Access

Parkgate Marsh can be accessed on minor roads just south-west of Neston. There are car-parking facilities and a public footpath runs along the edge of the marsh.

Site description

Situated on the northern side of the Dee Estuary, this saltmarsh attracts huge flocks of waders and wildfowl during high tides in winter. Water Shrews and Pygmy Shrews both become viewable during such periods as the water reaches the wall. The creeks provide ideal habitat for these species.

2 Formby Point, Merseyside **Grid ref: SD276082**

Key species

Red Squirrel.

Access

Formby Point is approximately halfway between Liverpool and Southport and is easily accessed from the A565 through Formby town. It is open all year round from dawn till dusk (except Christmas Day). Parking costs £3.30 for cars (in 2006). The overflow beach car park closes at 17:30 hrs, and between November and March all beach car parks close at 16:30 hrs. There is a disabled toilet opposite the main notice board.

Red Squirrel (*U. Iff*)

Site description

The National Trust property at Formby Point consists of 182 hectares of pinewoods, beaches and sand dunes and is an excellent site for **Red Squirrels.** Although in some eyes it could be seen as something of a wildlife park, the coniferous woodlands do provide a safe refuge for the reds from their marauding grey cousins. It is possible to buy bags of food with which to feed the animals: although this may not do much for their 'wild' animal tag, it does provide excellent viewing opportunities. Away from the Point, they can also be seen further up the coast at Ainsdale, and inland at Ruff Wood, near Ormskirk.

Other information

Natterjack Toads breed in some of the dune slack pools along this stretch of coastline.

3 Pennington Flash Country Park, Greater Manchester

Grid ref: SJ643991

Key species

Soprano Pipistrelle, Common Pipistrelle, Noctule, Daubenton's Bat, Brown Long-Eared Bat, Natterer's Bat.

Access

The main entrance is signposted from the A580 (East Lancashire Road). The park is managed by the Wigan Countryside Service and is fully accessible to the public.

Site description

A mix of habitats, with deciduous woodland, grassland and flashes (water-filled hollows). The main flash and surrounding pools provide the focus for the bat interest. Also, **Noctules** are easily visible over the field and tree lines next to the visitor centre.

4 Pennines/Peak District, South Yorkshire/ Derbyshire

Grid ref: SK068984 (Long Gutter Edge)

Key species

Mountain Hare.

Site description

In the Pennines and the Peak District, Mountain Hares are found on heather moors at altitudes of over 300m. They tend to favour the boulder-strewn slopes below scarps, and gritstone edges. One study suggests that there are about 500 hares present in the spring and about double this number in autumn after breeding. This may only just be a viable population.

Mountain Hares are quite easy to locate in the southern Pennines. A particularly good site is Long Gutter Edge (SK068984) near Torside Reservoir in Derbyshire. From junction 36a off the M1 follow the signs for Manchester along the A628. Woodhead Reservoir will be reached after about eight miles (13km). Immediately after the reservoir take a left turn along the B6105 towards Glossop. After about two miles (3km) there is a car park on the left. Park here and scan the scree slopes and moorland edges above the car park. Early mornings and late afternoons are best with up to 16 animals seen here on a regular basis in recent years. The car park is also quite good for **Stoats**.

Other good places to see Mountain Hares include Bleaklow Hill, and on moorland around Ladybower Reservoir.

5 Blacktoft Sands RSPB Reserve, East Yorkshire

Grid ref: SE842231

Key species

Water Shrew, Water Vole.

Access

Situated about ten miles north of Scunthorpe, this RSPB reserve can be accessed from the minor road off the A161 through Reedness and Ousefleet. The reserve is open daily (except Christmas Day) from 09:00–21:00 hrs or dusk. Access is free to RSPB members. Charges for non-members are: adults £3, children £1, concessions £2, family ticket £6.

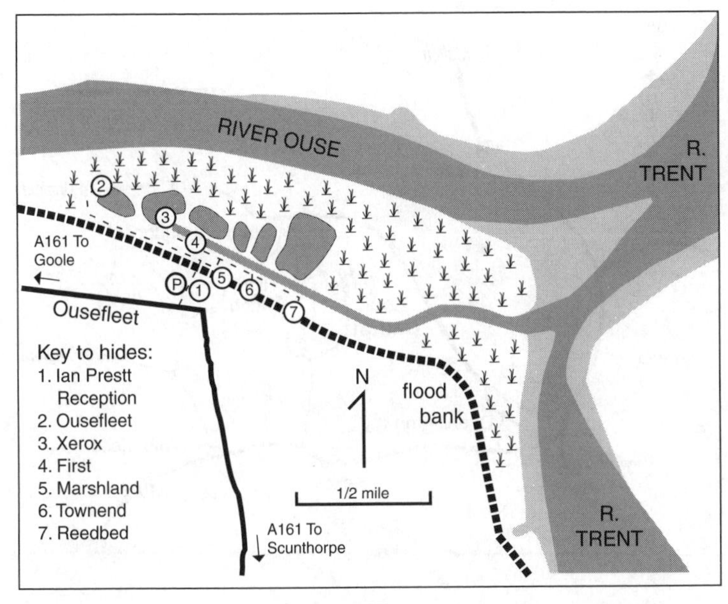

Site description

The reserve consists of 192 hectares of tidal reedbeds, saltmarsh, mudflats and brackish lagoons, much of which is suitable for the key species. There are six hides (five of which are suitable for wheelchair users) and a trail that runs around the reserve. Both **Water Shrews** and **Water Voles** are best located from hides overlooking the reedbeds, although many of the ditches are also worth looking at. Other mammals include **Brown Hares** and **Roe Deer**.

Other information

This reserve is excellent for wintering wildfowl and passage waders. Breeding birds include Avocets and Bearded Tits.

6 Low Catton, East Yorkshire Grid ref: SE704540

Key species

Natterer's Bat.

Access

Low Catton is approximately six miles (9.5km) east of York, and the church can be reached via the small minor road at the sharp turn in the village.

Site description

The village church in Low Catton hosts roosting **Natterer's Bats** during the summer and these can be watched leaving the roost within the church and hunting over the adjacent countryside.

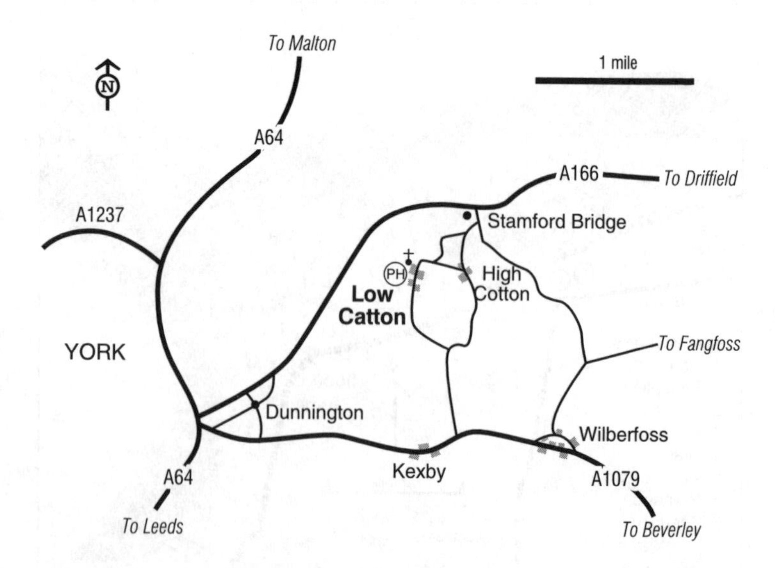

7 Boltby Forest, North Yorkshire

Grid ref: SE515830 (Sneck Yate car park)

Key species

Fallow Deer, Roe Deer.

Access

The forest is approximately five miles (8km) east of Thirsk and three miles (nearly 5km) north of the A170 at Sutton Bank. A good place to start is at Sneck Yate. This spot offers a small parking area at a minor crossroads. From Sutton Bank take the minor road north to Cold Kirby. At the next junction follow signs for Boltby and the parking area is a further three miles (5km) along this road. Alternatively, this woodland can be accessed via the Cleveland Way long-distance footpath that passes through the forest. The area lies on the western banks of the Hambleton Hills in the North Yorkshire Moors National Park.

Site description

Much of this large forest is managed by the Forestry Commission. It is dominated by coniferous woodland in the north and deciduous woodland to the south. **Fallow Deer** can be observed throughout the forest but are more likely to be encountered in the northern section. Other mammals that occur here include **Red Foxes**, **Badgers** and **Roe Deer**.

8 St Hilda's Church, Ellerburn, North Yorkshire

Grid ref: SE841842

Key species

Natterer's Bat, Whiskered Bat, Brown Long-eared Bat, pipistrelle species.

Access

Ellerburn, which contains a church and one other house lies a mile (1.6km) to the north of Thornton-le-Dale, three miles (nearly 5km) to the east of Pickering, and can be accessed via the minor road north from Thornton-le-Dale.

Site description

The 1,000-year-old Saxon church is perhaps Yorkshire's most important bat site being home to over half the county's resident species and including one of Britain's most important **Natterer's Bat** colonies. The St Hilda's Church population is the largest of the 24 Natterer's Bat colonies in North Yorkshire. **Natterer's** and **Brown Long-eared Bats** both roost in the church itself. They are viewed by certain churchgoers as unwelcome guests because of the irreparable damage they are causing to the upholstery. By April 2004, the congregation had fallen to just 11 regulars. **Whiskered Bats** and **pipistrelles** roost in the roof.

9 Wykeham Forest, North Yorkshire

Grid ref: Highwood Brow: SE943889
Raptor Watchpoint: SE936887

Key species

Brown Long-eared Bat, pipistrelles, Roe Deer.

Access

Wykeham Forest is approximately six miles (9.5km) west of Scarborough off the A170. It is signposted 'North Moor'. Two small car parks at Highwood Brow offer views over the Upper Derwent Valley and the North Yorkshire Moors National Park. The Highwood Brow car parks are located between Wykeham and North Moor immediately opposite a T-junction. The raptor watchpoint can be found by turning left at this junction, following the road for about one third of a mile, and the car park is on the right-hand side. This latter site offers fantastic views over the surrounding countryside and offers good wheelchair access.

Site description

This Forestry Commission woodland on the southern Tabular Hills of the North Yorkshire Moors is being used as a centre for research into 'Alternatives to Clearfell' forestry. Both **Brown Long-eared Bats** and **pipistrelles** use bat boxes that have been erected within the forest and can be seen hunting at dusk from the raptor watchpoint. The best chance of observing **Roe Deer, Badgers** and **Red Foxes** is at Highwood Brow where they are occasionally seen on the tracks and fields. These species have also been observed from the raptor watchpoint at daybreak.

Other information

Honey Buzzards may be observed from the raptor watchpoint between mid-May and late August.

10 North Riding Forest Park, North Yorkshire

Grid ref: SE775944

Key species

Roe Deer.

Access

This large area is easily accessed from Scarborough, Whitby or Pickering. There are many woodland trails and roads running through the forest.

Site description

The North Riding Forest Park is within the North Yorkshire Moors National Park and consists of 12,000 hectares of productive coniferous forest. Studies have indicated that the park has a population of 6.75 **Roe Deer** per km^2. Productive areas include the forest drives through Cropton Forest and north of Wardle Rigg.

Otter (*U. Iff*)

11 Upper Derwent, North Yorkshire

Grid ref: Whitby: NZ896108
Scarborough: TA040887

Key species

Otter.

Site description

An **Otter** reintroduction scheme has been running since 1990. During November 1997 there was a number of reports of **Otter** sightings in Whitby harbour and along the coast at Scarborough and up the coast to Saltburn. **Otters** are now visiting most of the coastal streams north of Scarborough.

12 Mount Grace Priory, North Yorkshire

Grid ref: SE448985

Key species

Stoat.

Access

Mount Grace Priory is just off the A19, 12 miles (19km) north of Thirsk. It is a National Trust property open to the public from 10:00–18:00 hrs.

Site description

One of the most famous sites for Stoats, this was the setting for the BBC's programme 'Stoats in the Priory'. The National Trust encourage the **Stoat** population as they help control the numbers of **Rabbits** on the property. The animals themselves have become accustomed to humans and can sometimes be seen attacking Rabbits in full view of visitors.

13 Seal Sands, Cleveland

Grid ref: NZ529258

Key species

Grey Seal, Common Seal.

Access

A number of tracks accessed from the A178 skirt Seal Sands.

Site description

The once extensive intertidal mudflats of Tees Bay once supported a colony of thousands of seals until the latter half of the nineteenth century, when they were displaced by industrial development. Only 150 hectares of the original Seal Sands remain. The return of the seals is thought to be their only recolonisation of a formerly degraded estuary in the British Isles. The breeding grounds within Tees Bay are on the last remaining significant area of mudflats in the Seaton Channel.

Grey Seal: Since the Tees Seals Project began in 1988, Grey Seal numbers have increased slowly from 18 to 27 individuals in 1996. Individuals have started to be spotted on the River Esk at Whitby. They have been observed in the harbour and occasionally three miles (nearly 5km) up river at Sleights Weir.

Common Seal: The maximum number of Common Seals hauling out at Seal Sands increased from 23 in 1989 to 73 in 2001.

14 Teesdale, County Durham

Grid ref: NY947252 (Field Studies Centre at Middleton-in-Teesdale)

Key species

Whiskered Bat, Brandt's Bat, Natterer's Bat, Noctule, Brown Long-eared Bat, Daubenton's Bat, Soprano Pipistrelle, Common Pipistrelle.

Access

The Middleton-in-Teesdale site is approximately eight miles (13km) north of Barnard Castle and is accessed via the B6277. Parking for the Whiskered Bat colony is available in the village. There is also a car park at High Force Waterfall and public footpaths run along both banks of the River Tees for most of its length between Middleton-in-Teesdale and High Force Waterfall.

Site description

Teesdale is nationally renowned for its plant fauna with several species endemic to the valley. The area, part of the North Pennines Special Protection Area, is characterised by sympathetically managed meadows, juniper-clad hillsides and scars of broken rocks that lead up to the heather moorland that cloak the higher plateaus. All eight bat species previously listed can be found in varying numbers between High Force Waterfall and Middleton-in-Teesdale in summer.

A large colony of **Whiskered Bats** breeds in the Field Studies Centre at Middleton-in-Teesdale. The centre is on Bridge Street (B6277) by the River Tees, and the bats can be watched at close range as they leave the roost and commence foraging. They are best watched from the riverside public footpath that runs east from the main road (B6277) along the River Tees on the north bank. You only need to go a few yards through the gate from the road here as the roost is in the building immediately to the north. In midsummer the bats emerge well before dark and from the vantage point described above they can be watched flying overhead as they head out to feed over the Tees or to forage around the trees between the centre and the river. They are difficult to watch from the bridge as they feed over the river, as they tend to stay close to the well-vegetated north bank.

Smaller numbers of **Daubenton's Bats** also hunt over this stretch of the River Tees, whilst usually several **Noctules** can be watched hunting over the meadow on the opposite side of the road from the centre. Further east, good numbers of **Daubenton's Bats**, **Noctules** and **pipistrelles** may be observed foraging from Egglestone Abbey Bridge (NY996232).

In recent summers small numbers of **Leisler's Bats** and **Serotines** have started to be recorded from Teesdale. This would seem to indicate post-breeding dispersal into the area. The **Nathusius's Pipistrelle** has also been recorded recently, with single records from Cotherstone and from near Bowes. Further to the east at least four Nathusius's Pipistrelles were observed recently by members of the Durham Bat Group, feeding over Whitworth Hall Lake just north-west of Spennymoor (NZ234347).

15 Raby Castle, County Durham

Grid ref: NZ128217

Key species

Whiskered Bat.

Access

Raby Castle is just north of Staindrop in southern County Durham. The castle closes each winter but during summer it is open daily (except Saturdays) during July and August, and on Wednesdays and Sundays in May, June and September. The park and gardens are open daily (except Saturdays) from May until the end of September. In 2006, the admission charges to the park and gardens only were: adults £4, children £2.50, concessions £3.50. For further details of opening times and admission prices refer to www.rabycastle.com.

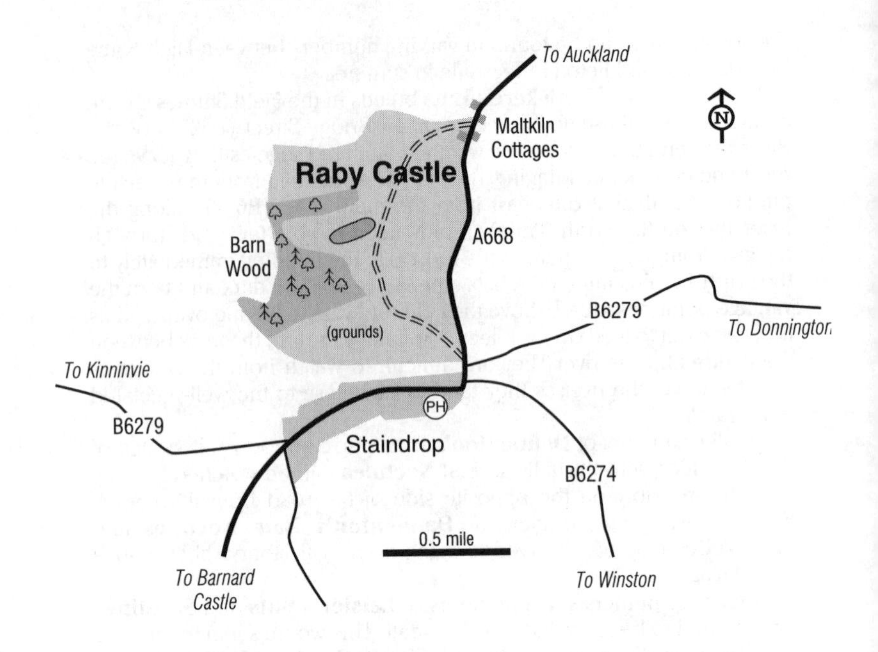

Site description

Raby Castle is one of England's largest and most impressive medieval castles and is an extremely good site at which to see **Whiskered Bats**. The bats may be seen around the main gate but the presence of **Brown Long-eared Bats**, **Natterer's Bats** and **pipistrelles** here can make identification difficult. The 80-hectare deer park that surrounds the castle is home to herds of both **Red Deer** and **Fallow Deer.**

16 Malton Nature Reserve and Picnic Site, County Durham

Grid ref: NZ176464

Key species

Brandt's Bat, Noctule, Daubenton's Bat, American Mink.

Access

Malton Nature Reserve is just off the A691 Durham to Consett road, approximately half a mile south-east of Lanchester. The picnic area is signposted and the reserve can be accessed on foot via Officials Terrace or via the Railway Path. There is a circular walk around the reserve of around a mile (1.6km). There is free parking as well as picnic tables and benches. The riverside path to the 'Water's Meetings' and picnic fields are accessible by wheelchair, although the riverside path is uneven in places and becomes muddy after periods of wet weather.

Site description

This 5.7-hectare nature reserve contains a wide variety of habitats and is managed to show examples of succession on colliery shales, from lichen heath to oak woodland. Malton Picnic Site is a good site at which to watch **Brandt's Bats**. They may be seen hunting over the River Browney at this site. Other bats that can be seen on the reserve include **Noctule**, **pipistrelles** and **Daubenton's Bat**. The latter can be seen foraging with **Brandt's Bats** over the river. Other species present include **Roe Deer, Red Foxes** and **Badgers**, while **Otters** and **American Mink** are occasionally seen.

17 Hamsterley Forest, County Durham

Grid ref: NZ092312

Key species

Brandt's Bat, Natterer's Bat, Whiskered Bat, Water Shrew.

Access

Hamsterley Forest is between Weardale and Teesdale. The main entrance at South Bedburn is reached via Hamsterley village or Wolsingham. It is also signposted from the A68, three miles (nearly 5km) north of West Auckland. Facilities include a visitors' centre and forest shop, waymarked walks and cycle routes, a four-mile (6km) forest drive (£2 per car) and an adventure play area. The forest is open throughout the year from dawn to dusk and the visitors' centre is open daily: weekdays 10:00–16:00 hrs; weekends 11:00–17:00 hrs, from Easter to October (as well as December) and at weekends only in November. There is wheelchair access to the visitors' centre and main picnic areas, and along the riverside footpath.

Site description

Hamsterley Forest is a 2,000-hectare Forestry Commission managed woodland. There are year-round harvesting operations with signs notifying visitors of areas of restricted access. While most of the forest consists of conifers, around 25 hectares of deciduous woodland can be found along the Bedburn Brook, while a section of Beech woodland at an altitude of over 350m is reputed to be England's highest example of this habitat.

A large bat-box scheme in Hamsterley Forest is run by the Durham Bat Group. In total, eight bat species can be found within the woodland. The forest is host to the largest **Natterer's Bat** breeding colony in County Durham. In 2003 a maternity colony of 68 individuals was found in just one bat box. Other bat species present include **Brandt's Bat** and **Whiskered Bat**, with several commoner species also present.

Other mammal species include **Roe Deer**, **Badgers** and **Grey Squirrels**. The forest was once a stronghold of Red Squirrels, but unfortunately their North American cousins squeezed this species out very recently. **Water Shrews** and **Otters** may be observed with luck and patience along the becks that run through the forest.

Other information

At the western end of the forest lies Neighbour Moor where you can see species such as Goshawk and Common Buzzard from a raptor watchpoint.

18 Burrs Country Park, Bury, Lancashire

Grid ref: SD799127

Key species

Soprano Pipistrelle, Common Pipistrelle, Daubenton's Bat.

Access

The entrance and car park are at the end of Woodhill Road, signposted from the B6214 Brandlesholme Road approximately one mile (1.6km) from Bury town centre. A path from the car park leads to the canoe pool and along the River Irwell. Another path leading around the back of the activity centre takes you to a circular walk along the river and past two old millponds.

Site description

Two millponds, the canoe pool and the river provide the focus for the bat interest. The park, which is managed by the Bury Countryside Service, is fully accessible to the public. **Noctules**, **Whiskered** or **Brandt's Bats** and **Brown Long-Eared Bats** have been recorded nearby.

19 Jumbles Reservoir and Country Park, Bolton, Lancashire

Grid ref: SD736140

Key species

Soprano Pipistrelle, Common Pipistrelle, Daubenton's Bat, Noctule, Brown Long-Eared Bat.

Access

The entrance to the reservoir is on the A676 just north of Bolton. The reservoir and country park are signposted.

Site description

Deciduous woodland. The reservoir and the open areas along the path provide the focus for the bat interest. The reservoir, which is managed by United Utilities, is fully accessible to the public. However the toilets and cafe are shut during the evening.

20 Yarrow Valley Park, Lancashire

Grid ref: SD570152

Key species

Soprano Pipistrelle, Common Pipistrelle, Noctule, Whiskered Bat, Daubenton's Bat.

Access

Yarrow Valley Park is located between Chorley and Coppull and is accessed off the B5251.

Site description

Good numbers of **Whiskered Bats** and **Daubenton's Bats** can regularly be seen from the bridge across a stream at the southern end of the main reservoir.

21 Leighton Moss RSPB Reserve, Lancashire

Grid ref: SD481749

Key species

Water Shrew, Noctule, Red Deer.

Access

This well-known RSPB reserve is two miles (3km) north of Carnforth and is signposted from the A6 north of junction 35 of the M6. The reserve is open daily (except Christmas Day) from 09:00 to 21:00 hrs or dusk. The visitors' centre is open daily from 09:30 to 17:00 hrs (16:30 hrs from November to January). Several nature trails criss-cross the reserve and some are suitable for wheelchairs. There are five hides, four of which have wheelchair access. Parking, toilets and a tearoom are all available. Admission to the hides and nature trails is free to members and those who arrive by public transport or bicycle. Otherwise, charges are: adults £4.50, children £1, family £9, concessions £3. Guided walks are available throughout the year, some of which are specifically to look for Red Deer.

Site description

The reserve incorporates a range of habitats but is most famous for its reedbed, the largest in north-west England, and bird life including Bitterns, Marsh Harriers and Bearded Tits. It also contains freshwater scrapes, shallow meres and fringing woodland and scrub all of which attract a range of wildlife. This is an excellent site for **Water Shrews** and the species is common here. Up-to-date information regarding recent sightings may be obtained from the reception area, but the best chances of seeing one is probably to stand on one of the footpaths that run straight through the reed-beds, as the shrews cross these paths on a regular basis. A particularly good spot is the footpath leading out to the Grizedale Hide.

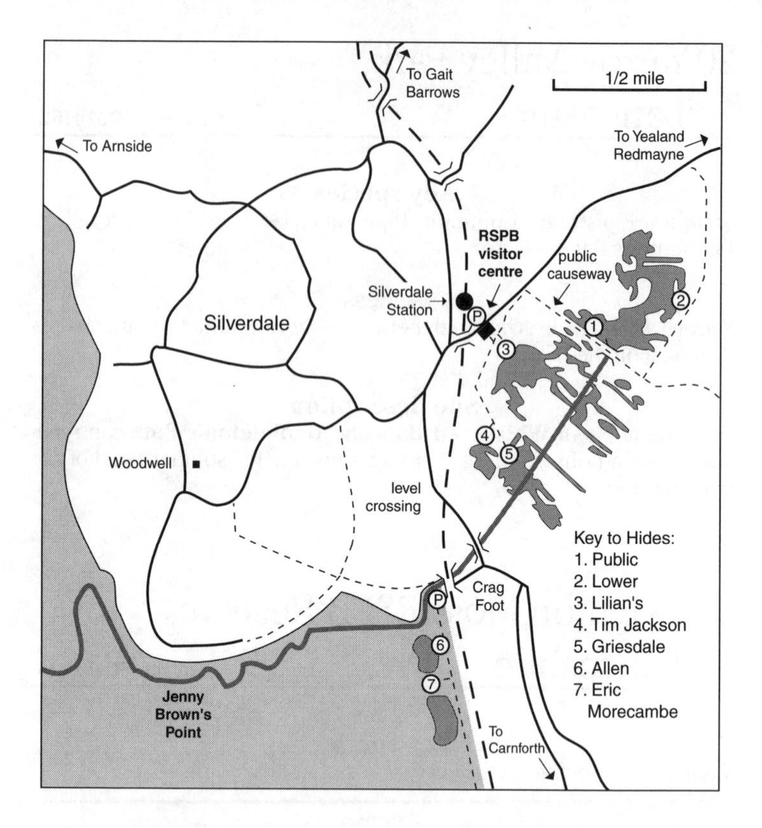

Noctules are quite numerous at Leighton Moss and are easy to observe on a summer evening as they hawk insects over the reserve. Winter nests of **Harvest Mice** have been found in Reed Canary Grass on the reserve. **Roe Deer** and **Red Deer** also occur. Nearby, **Whiskered Bats** are common in the Silverdale area (SD463749).

22 Grizedale Forest Park, Cumbria

Grid ref: SD338932 (Bogle Crag)

Key species

Red Squirrel, Red Deer.

Access

Grizedale Forest Park, managed by the Forestry Commission, is sandwiched between Lake Windermere and Coniston Water. Once in the vicinity, the park is well signposted from most directions. Bogle Crag, midway between the Forestry Commission visitors' centre and the village of Satterthwaite, offers a choice of three circular walks through ancient broadleaf woodland. Other good starting points include the Moor Top car park (SD343965) that has good views to the south and west and is the start of a number of trails.

Site description

As well as extensive tracts of coniferous plantations, Grizedale Forest Park also holds 250 hectares of upland oak wood. This area is one of the best places to see **Red Squirrels** in Cumbria, and with a bit of patience they can usually be seen from the trails that radiate into the woodland from the car park at Bogle Crag. **Red Deer** are also numerous and are best looked for away from the woodland. The forest edges looking west from the minor road opposite Bogle Crag car park can be productive, particularly early in the morning or late in the evening. The hillsides that can be viewed from Moor Top car park may also produce sightings.

23 River Derwent and Bassenthwaite Lake, Cumbria

Grid ref: NY215294 (Bassenthwaite Lake)

Key species

Otter.

Access

Bassenthwaite Lake is adjacent to the A66 just north of Kendal. The River Derwent continues from the lake to the sea at Workington.

Site description

The River Derwent and Bassenthwaite Lake provide a wide range of suitable conditions for **Otters**. The nearby River Eden also supports a healthy Otter population.

24 The Curraghs, Ballaugh, Isle of Man

Grid ref: SC364947

Key species

Red-necked Wallaby.

Access

Ballaugh is on the north-west side of the Isle of Man and is easily accessed via main roads from Douglas. The Curraghs woodland is just two miles (3km) north-east of Ballaugh village.

Site description

During the 1990s several Red-necked Wallabies escaped from the nearby Curraghs Wildlife Park and the resulting colony has since flourished. The locals welcome their presence and their numbers have reached at least 40 individuals. The best times to view the animals are at dawn and dusk along any of the footpaths and lanes in the area, particularly Penny Holding's Lane.

25 Kielder Forest, Northumberland

Grid ref: NY632934

Key species

Red Squirrel, Roe Deer, Otter, Brown Long-eared Bat, Common Pipistrelle.

Access

To reach the heart of Kielder Forest leave the A69 turning north onto the A6079 to Chollerford. Follow the B6320 to Bellingham and then the C200 to Kielder. If travelling from the north or south along the A68, take the B6320 and follow tourist signs for Kielder Water and Forest. There are excellent visitor facilities including a visitors' centre at Leaplish. Trails and tracks lead deep into the forest.

Site description

There are 150 million trees in Kielder Forest spanning an area of over 500km^2 from County Durham to the Scottish Borders. It is England's largest forest and has been almost entirely planted by humans. While it is composed of mainly exotic species such as spruce, it constitutes an ideal home for **Red Squirrels** and currently harbours around 12,000 individuals, about 75% of the English population. It is possible to view the squirrels in much of the forest, but good areas include Spadeadam Forest (designated as a Red Squirrel conservation area) and Leaplish. Kielder Forest supports good numbers of **Roe Deer** and these can be seen in most parts of the Forestry Commission woodlands. **Otters** use all the major waterways in the forest.

Kielder Castle (NY631933), located in the village of Kielder, is a good site for watching various species of bat. **Brown Long-eared Bats** and **Common Pipistrelles** roost within the roof space of the castle and may be observed hunting around the castle and surrounding environment throughout the summer. Other bat species that have been recorded within Kielder Forest include **Whiskered Bats**, **Brandt's Bats**, **Natterer's Bats**, **Daubenton's Bats** and **Noctules**.

Brown Long-eared Bat (*R. Lindsay*)

26 Rothley Lakes, Northumberland

Grid ref: NZ041907

Key species

Otter, Red Squirrel.

Access

The reserve can be accessed north off the B6342 between Cambo and Rothbury through a gate onto a single-track road. After a further 365m (400 yards) pull into another gate on the right and park on the track. Although some of the shoreline is off-limits much of the lake can be easily viewed. Please do not cross into the adjoining private land.

Site description

Capability Brown created the lakes at Rothley in the mid-eighteenth century, and they occupy a glacial meltwater channel. Although Rothley Lakes are not strictly part of this 20-hectare Northumberland Wildlife Trust reserve, the shores of the artificial lakes form the boundary allowing viewing of visiting **Otters**. The rest of the reserve consists of a disused railway cutting and a mixed plantation. Whilst Otters are only regular visitors, **Red Squirrels** are common residents within the woodland as well as in Broomfield Fell woodland just to the east. Bat boxes have recently been erected on some of the trees.

27 Shibdon Pond LNR, County Durham

Grid ref: NZ195628

Key species

Noctule.

Access

Shibdon Pond is just to the south-east of Newcastle-upon-Tyne and can be accessed from the A1 at the Blaydon exit. The B6317 skirts the reserve's southern boundary. Shibdon Pond has a network of footpaths and boardwalks that take you through the various habitats.

Site description

This 14-hectare Local Nature Reserve and Site of Special Scientific Interest (SSSI) has at its centrepiece Shibdon Pond, created by subsidence from mining activity. The depression has filled in with water and has created an urban marsh and wetland. During the summer, **Noctules** may be seen hawking insects above the pond and over the trees. **Daubenton's Bat** has also been recorded on the reserve but away from the main pond. **Common Pipistrelles** are plentiful.

28 Chopwell Woodland Park, County Durham **Grid ref: NZ137586**

Key species

Otter, Badger.

Access

The entrance is off the B6315 at the southern end of High Spen village. There is a car park on site.

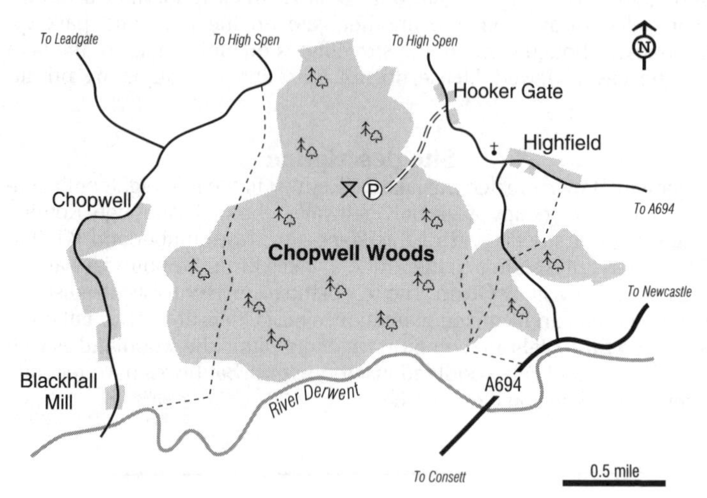

Site description

This mixed woodland covers 360 hectares and abuts the River Derwent on its southern boundary. **Otters** occur on the banks of the River Derwent and may be seen with luck at dawn and dusk. Other species that may be seen in Tyne and Wear's largest woodland include **Badgers**, **Roe Deer** and **Red Foxes**.

29 Annitsford Pond Local Nature Reserve, Northumberland **Grid ref: NZ267743**

Key species

Water Vole.

Access

Annitsford Pond is half a mile west of Seaton Burn along the A189 approximately six miles (9.5km) north of Newcastle. From Annitsford Front Street turn into Harrison Court, park and follow the footpath running parallel to the stream called Sandy's Letch. The reserve enjoys open access.

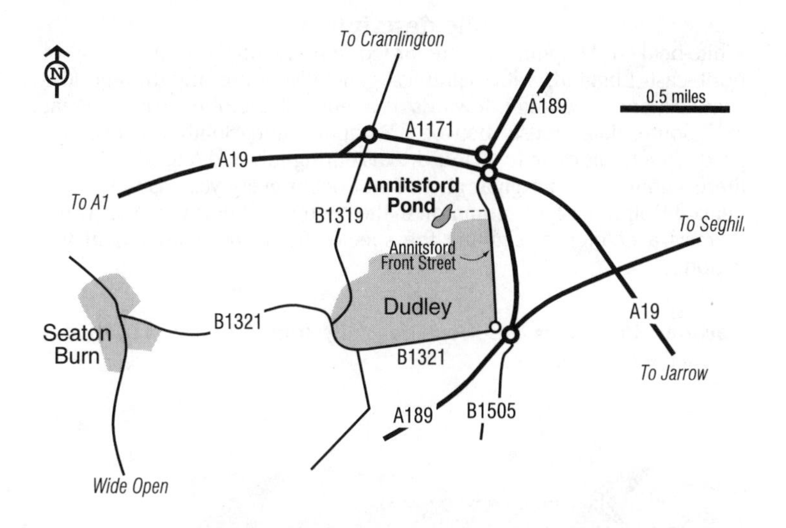

Site description

This is a small subsidence pond that has been managed jointly by North Tyneside Council and the Northumberland Wildlife Trust since 1984. The reserve consists of a variety of habitats but the **Water Voles** are restricted to the areas of open water with tall, emergent vegetation. This is an excellent site to view this species.

30 Whitley Bay, Northumberland

Grid ref: Lizard Point (Souter Lighthouse): NZ410641. St Mary's Lighthouse (St Mary's Island): NZ352754

Key species

White-beaked Dolphin, Harbour Porpoise.

Access

Souter Lighthouse on Lizard Point is a National Trust property just north of Whitburn. There is a free car park at the point. Alternatively, it can be accessed north along a public footpath from Whitburn Coastal Park. This coastal path also runs south to Souter Point.

St Mary's Island and Lighthouse can be found just north of Whitley and can be accessed via a drivable track off the A193. There is a car park and the track continues to the lighthouse. The island can only be accessed at low tide.

Site description

White-beaked Dolphins are recorded infrequently off the coast of north-east England, with sightings generally occurring through the summer and declining towards the end of October. Souter Point and Souter Lighthouse, between Whitburn and South Shields, can produce schools of up to 40 individuals during July and August, although these sightings are irregular and do not occur every year. As the White-beaked Dolphin is fairly common in the central and northern North Sea, there is a chance of sighting the species from any headland in the region.

Harbour Porpoises are seen irregularly from St Mary's Lighthouse during the autumn.

White-beaked Dolphin (*Dan Brown*)

31 Carlisle Park, Morpeth, Northumberland

Grid ref: NZ198857

Key species

Common Pipistrelle, Soprano Pipistrelle, Daubenton's Bat.

Access

Park by Morpeth Leisure Centre, south of Morpeth Town Hall, and access the park on foot over Elliott Bridge.

Site description

This is a Local Nature Reserve containing areas of ancient semi-natural woodland. The River Wansbeck, over which **Daubenton's Bats** can be seen feeding, forms the northern boundary of the park. **Whiskered Bats** and **Natterer's Bats** have also been recorded.

32 Corbridge, Northumberland

Grid ref: NY988641

Key species

Pipistrelles, Natterer's Bat, Noctule.

Access

There is a car park on the south side of the B6321 bridge on the south edge of the town. A public footpath follows the River Tyne in both directions.

Site description

The public footpath that runs east along the river passes through woodland where **pipistrelles** and **Natterer's Bats** feed over the footpath adjacent to the playing fields. **Noctules** can be observed over the trees or playing fields.

33 Rothbury, Northumberland

Grid ref: NU059016

Key species

Daubenton's Bat, Natterer's Bat, pipistrelles, Noctule.

Access

Park in the centre of Rothbury or in the riverside car park. Walk back to the river bridge or paths along the river.

Site description

The key species may be seen hunting over the river and nearby. **Water Voles** and **Otters** may also be seen within 100m of the bridge.

34 Farne Islands, Northumberland

Grid ref: NU242369

Key species

Grey Seal.

Access

The Farne Islands lie two miles (3km) off the Northumberland coast and can be accessed by boat from the end of March to the end of September. Boats leave for the Farnes from Seahouses on the mainland. Further information can be found at www.farne-islands.com or www.nationaltrust.org.uk.

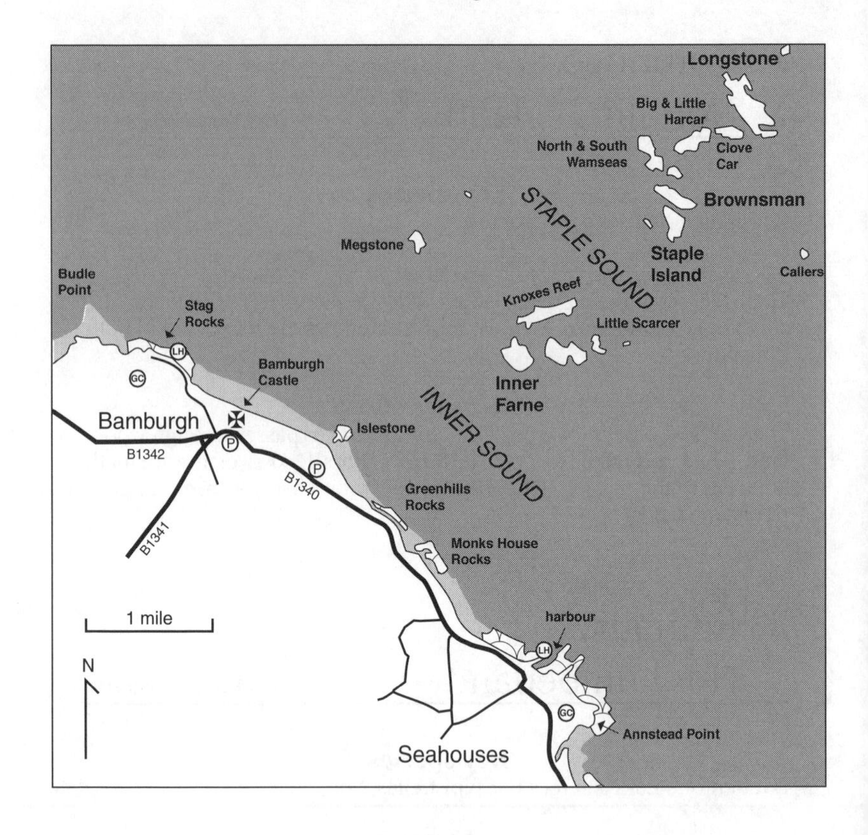

Site description

At the start of 2001, the Farne Islands supported about 4,000 **Grey Seals** and these can be easily observed during trips to the islands. Grey Seals also occur in good numbers along the north Northumberland and Berwickshire coastlines.

Other information

The Farne Islands are one of Europe's most important seabird colonies, with more than 20 different species nesting on

WALES

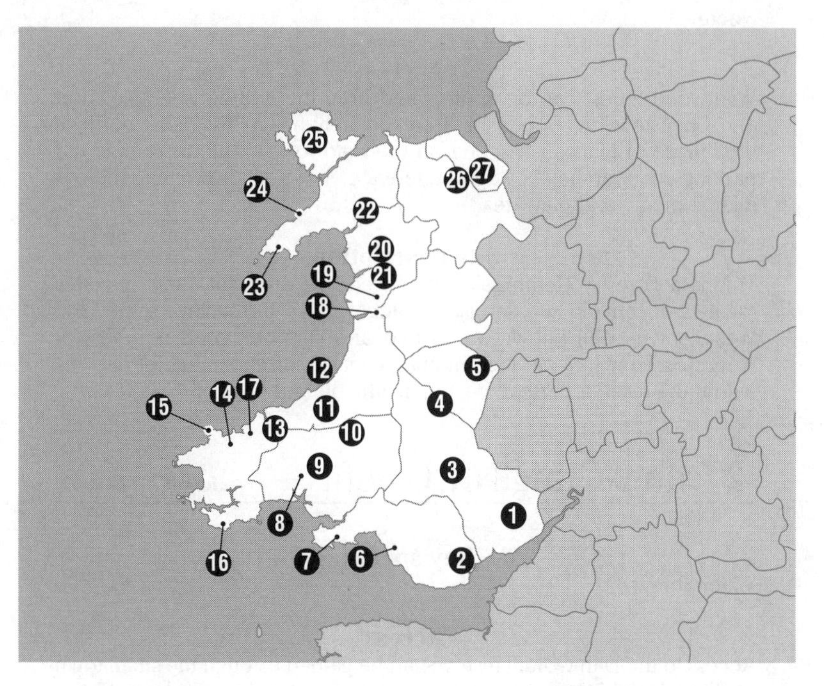

1 Wentwood Forest
2 Dan-y-Graig NR
3 Brecon Beacons National Park
4 The Lake
5 Radnor Woods
6 Afan Forest Park
7 Worm's Head and Burry Holms, Gower Peninsula
8 River Tywi/Towy
9 Brechfa Forest
10 Crychan
11 Tregaron Bog/Cors Caron NNR
12 Cardigan Bay
13 Teifi Marshes Nature Reserve
14 North Pembrokeshire woodlands
15 Strumble Head and the Pembrokeshire coast
16 Stackpole Estate
17 Pencelly/Pengelli Forest
18 Dyfi
19 Tan y Coed
20 Coed y Brenin
21 Penmaenuchaf Hall Hotel
22 Snowdonia
23 The Lleyn Peninsula and Bardsey Island
24 Coastal Gwynedd
25 Anglesey
26 Moel Famau
27 Nercwys

1 Wentwood Forest, Gwent **Grid ref: ST424945**

Key species

Noctule.

Access

Wentwood Forest can be approached from the A48 between Newport and Chepstow. The woods are approximately five miles (8km) north of the village of Llanvair Discoed on the road to Usk. There is ample car parking either at the above grid reference or at ST418946, where there is a large picnic and play area.

Site description

This is a Forestry Commission managed site of about 1,000 hectares located on an old red sandstone ridge above the valley of the Usk. Wentwood is well known for its bats and supports good numbers of **Noctules**. They may be seen hunting over the diverse range of habitats within the reserve, particularly the heathland and boggy areas.

2 Dan-y-Graig NR, Gwent **Grid ref: ST235905**

Key species

Water Shrew.

Access

Access to the Dan-y-Graig reserve can be gained through the small town of Risca. Head into the centre of town before turning left down Dan-y-Graig Road. Cars can be parked under the bypass. Walk uphill to the left of Dan-y-Graig Cottages and through a gate to the reserve.

Site description

This is a small 2-hectare nature reserve managed by the Gwent Wildlife Trust. It consists of calcareous grassland, broadleaved woodland and a pond. The pond was originally built to provide water for the copper works that stood on the site of the adjacent brickyard. It is now the home of a healthy population of **Water Shrews**.

3 Brecon Beacons National Park, Powys **Grid ref: SN819285 (Usk Reservoir)**

Key species

Polecat, Red Deer.

Access

From the A40 between Brecon and Llandovery turn off at Trecastle and follow signs to Usk reservoir. Glasfynydd Wood surrounds Usk Reservoir.

Site description

This is obviously a very large area but it supports a sizeable population of **Polecats**. One area to try is the Usk reservoir and surrounding coniferous woodlands. The plantation, named Glasfynydd, is owned by the Forestry Commission and provides ideal habitat for Polecats. It also offers sanctuary for the **Red Deer** that graze along the woodland edge in the shadow of the Black Mountain.

4 The Lake, Powys

Grid ref: SO062605

Key species

Daubenton's Bat, Noctule, pipistrelles.

Access

The Lake is on the southern outskirts of the small town of Llandrindod Wells in mid-Powys. Viewing is simple as the lake is sandwiched between Princes Avenue and Grosvenor Road, both of which are reached via the A483 that runs through the town centre.

Site description

All of the target species are easy to observe at dusk between May and September.

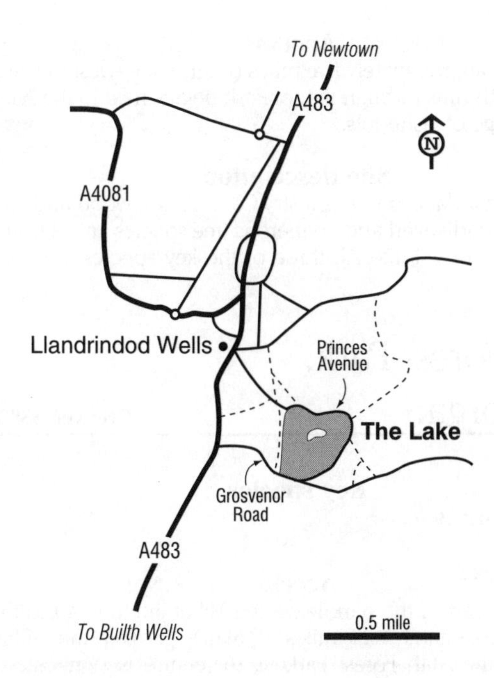

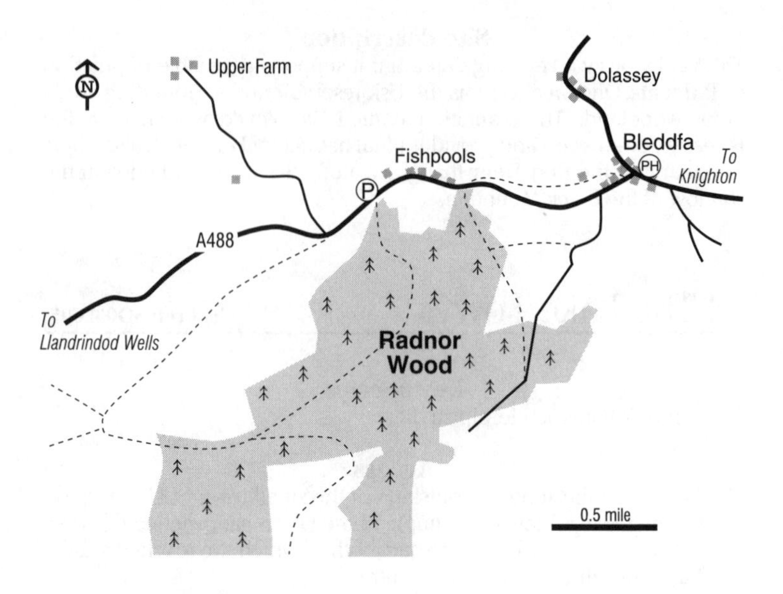

5 Radnor Wood, Powys

Grid ref: SO188683

Key species

Badger, Red Fox, Roe Deer.

Access

Radnor Wood is approximately five miles (8km) south-west of Knighton. A public footpath runs through the forest. It begins next to the A488, just west of the village of Fishpools.

Site description

This large and easily accessible Forestry Commission woodland incorporates a mix of broadleaved and coniferous tree species and is home to a diverse selection of wildlife. All three of the key species are common here.

6 Afan Forest Park, Glamorgan

Grid ref: SS820950

Key species

Fallow Deer, Brown Hare.

Access

To get into the heart of the park leave the M4 at junction 40, follow the A4107 for approximately six miles (9.5km), passing the village of Cwmafan, and the Afan Forest Park visitor centre is signposted from here. The visitors' centre has pay-and-display parking facilities, toilets (including disabled), a café and a shop, and there are numerous trails through the forest. An alternative access point is Rhyslyn car park (SS800943). Here there is a large car park with picnic benches, together with views across the Afan Valley. To access Rhyslyn, turn down the B4287 towards Pontrhydyfen and take a sharp right turn just over the bridge as the road bears left.

Site description

Owned and managed by the Forestry Commission, Afan Forest Park is home to a herd of **Fallow Deer** that may be observed browsing amongst the higher conifer plantations. The White Walk along the river and up the hill offers particularly good views of the area and gets in amongst the plantations. **Brown Hares** have colonised the area relatively recently and numbers are now on the increase. Look for this species in suitable habitat, especially in grassy areas and amongst the young plantations

7 Worm's Head and Burry Holms, Gower Peninsula, Glamorgan

Grid ref: SS402873 (Worm's Head) SS399926 (Burry Holms)

Key species

Harbour Porpoise.

Access

The Gower Peninsula lies west of Swansea and is accessed via the A4118. Worm's Head is reached by taking the B4247 a couple of miles prior to reaching Port Eynon. There is a car park and National Trust visitor centre in Rhossili and a public footpath leads down to Worm's Head. At the northern end of the bay there is a car park at Broughton Burrows with a public footpath leading around the coast to Burry Holms.

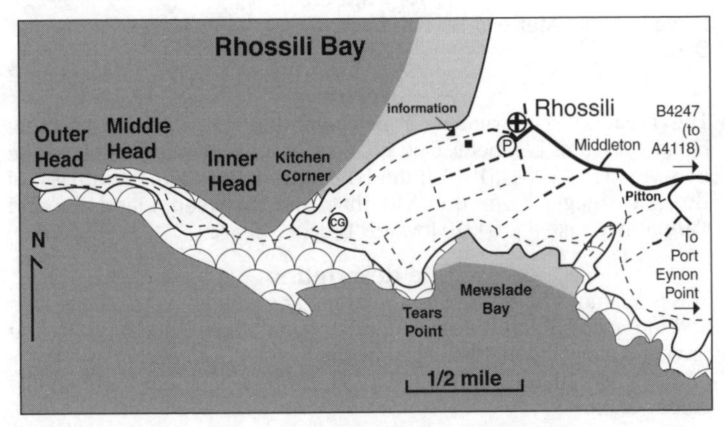

Site description

The Gower Peninsula was the first site in Britain to be designated as an Area of Outstanding Natural Beauty. Worm's Head and Burry Holms can provide sightings of **Harbour Porpoises** during spring. Both headlands/islets lie at the west end of the peninsula at either end of Rhossili Bay. The village of Rhossili offers dramatic views across the bay.

8 River Tywi/Towy, Carmarthenshire

Grid ref: SN633230 (Llandeilo)
SN369100 (Ferryside)

Key species

Otter.

Access

This extensive stretch of river is easily accessed at many points as numerous bridges cross it and stretches of public footpath run alongside the river at various stages. There is a car park at Ferryside from where it is possible to cross the railway to view much of the river mouth, and to walk a certain distance upstream.

Site description

This is one of the best rivers in Wales for **Otters**. There are abundant signs of Otters, and they can be observed along the river bank or in the water. The water quality is generally good and there is an ample supply of food. There are suitable lying-up areas on the banks, and although few breeding sites are known, cubs have been observed. Any point downstream of Llandeilo is worth exploring, although where the river widens towards the sea at the village of Ferryside may offer the best opportunity as animals are regularly observed at this point.

9 Brechfa Forest, Carmarthenshire

Grid ref: SN530340 (central point)

Key species

Polecat, Pine Marten, Brown Hare, Water Vole, Hazel Dormouse, Badger.

Access

The area is easily accessed between the towns of Carmarthen, Llandovery and Lampeter. One good place to start searching is in the Byrgwm area (SN544316), on the B4310 two miles (3km) north-east of Brechfa village. From the A40 that runs between Llandeilo and Carmarthen take the B4310 towards Brechfa village.

Site description

This is a famous forest, used as a royal hunting ground in the Middle Ages and to provide South Wales with timber during the twentieth century. At present it is owned and managed by the Forestry Commission as a multipurpose woodland for timber production, recreation and wildlife conservation.

Polecat: Although Polecats are always difficult to see, they inhabit Brechfa Forest in good numbers. There are many tracks into the forest and these are worth walking with a spotlight as dusk falls.

Water Vole: Water Voles are only patchily distributed throughout Brechfa. However a good place to search is at the Abernant picnic site (SN586378) which has car parking next to the River Melinddwr, where a population of the voles breeds along its banks. The picnic site is located a short distance south of Rhydcymerau on the B4337. The B4337 is reached from either Llanybydder or off the A482 that runs between Lampeter and Llandovery.

Hazel Dormouse: Hazel Dormice may be located with a lot of hard work and a great deal of luck in Brechfa. A good place to search is at Abergorlech picnic site (SN586337). From the A482 between Lampeter and Llandovery, turn off for Llansawel along the B4302. Once in Llansawel follow signs for Abergorlech. **Badgers** also occur in the woods around Abergorlech.

Brown Hare: Brown Hares may be observed in the grassy areas, where the tree cover is restricted or reduced. The Abernant picnic site is a good spot.

Pine Marten: Although their existence has not been confirmed within the forest they almost certainly occur here, although probably at a very low density.

Other species that may be encountered include **Otters**, along the River Melinddwr, **Red Foxes** and a number of common **bat** species.

10 Crychan, Powys

Grid ref: SN798378

Key species

Polecat, Roe Deer, Hazel Dormouse.

Access

From Llandovery, take the A483 north for six miles (9.5km) before taking an unclassified road on the right, by the Glan Bran public house, for three miles (nearly 5km) until you arrive at a car park. Alternatively, Halfway Forest (SN835330) has roadside parking at the start of the 'Halfway Health Walk', which follows the River Nant y Dresglen, and is found by taking the A40 for five miles (8km) east out of Llandovery (towards Brecon): the Forestry Commission car parking area is signposted off here.

Site description

This is an extremely large Forestry Commission forest, around 4,000 hectares, to the north of the market town of Llandovery. As well as the vast swathes of conifers there is also heathland, bogs and streams, as well as some excellent vantage points. This easily accessible forest supports **Polecats** as well as sizeable populations of many other mammal species including **Roe Deer**, **Hazel Dormice**, **Badgers**, **Brown Hares**, **Red Foxes** and a number of common **bat** species.

11 Tregaron Bog/Cors Caron NNR, Ceredigion

Grid ref: SN693630

Key species

Polecat.

Access

Tregaron Bog lies about three miles (nearly 5km) north of the village of Tregaron and is reached on the B4343. There is a car park and observation tower open to the public and most of the reserve is open at all times, although some parts require a permit obtained from the Countryside Council for Wales warden. See www.ccw.gov.uk for further details.

Site description

Tregaron Bog is today the best remaining example of a raised bog in England and Wales. The habitats within the reserve include bog, wet heath around the edges and rough grassland with patches of willow carr. The bog used to be one of the **Polecats** last remaining strongholds when they were persecuted to local extinction throughout much of Britain. They remain common in the area and, with luck and no little patience, they are sometimes seen from the tower hide at dusk. Other mammals on the reserve include **Otters**, **Water Voles** and **Water Shrews**.

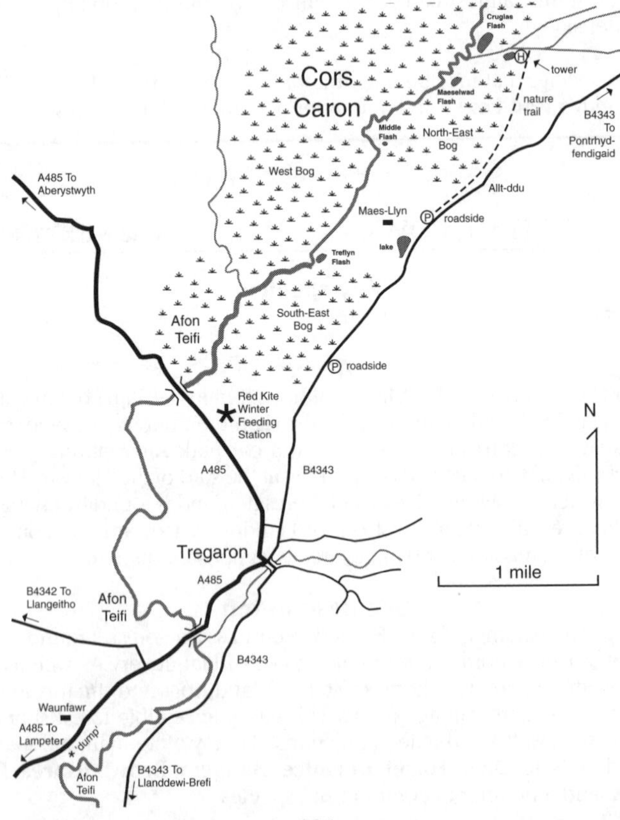

12 Cardigan Bay, Ceredigion

Grid ref: SN177461 (Cardigan town)

Key species

Bottle-nosed Dolphin.

Access

The A487 runs along the coast and provides access to any of the points mentioned below.

Site description

The Bay is a Special Area of Conservation. The resident population of Bottle-nosed Dolphins is centred between the towns of Cardigan and New Quay. They can regularly be seen from New Quay Head (SN385605) and around Cardigan Island (SN160516) in the Teifi Estuary. Groups consist of up to 40 animals. During the summer months they tend to move into a wider range of habitats and can be seen south to the north Pembrokeshire coast, and north to the Dovey Estuary and off Aberystwyth. However, they are generally easier to see and occur in larger numbers during the summer.

13 Teifi Marshes Nature Reserve, Ceredigion

Grid ref: SN186450

Key species

Water Shrew, Water Vole, Otter, Badger.

Access

Access to this Wildlife Trust for South and West Wales reserve can be on foot from Cardigan town along the river bank immediately upstream of the bypass bridge. By road, the Welsh Wildlife Centre is well signposted from Cardigan. Take the A478 south towards Tenby and after 1.5 miles (2km) turn left to Cilgerran and the centre is accessed via a long drive from the village. The centre is open daily from 10:30 to 17:00 hrs from April to October. The centre is closed in winter but the reserve remains open.

Site description

This 264-hectare nature reserve contains a diverse selection of wetland habitats along the River Teifi. A network of boardwalks passes through the marshes to a hide. With patience, both **Water Shrews** and **Water Voles** can be observed from the boardwalks. It is possible to watch **Otters** from the comfort of the centre, as there is a direct microcamera link from an active holt to television screens in the centre. **Badgers** come out at dusk to feed outside the centre, where food is provided for them.

14 North Pembrokeshire woodlands

Grid ref: SN046345

Key species

Barbastelle.

Access

To access these woodlands take the B4313 south-west from Fishguard for five miles (8km) before taking a left turning to Pontfaen. Go through the village for another two miles (3km) along the river valley and park in the picnic area, where there are also public toilets. There is a nature trail in the area although the woods can be viewed from the picnic area or from along the road.

Site description

This 316-hectare site is a complex of old sessile oak woodlands at that habitat's western limit within the UK. As well as the broadleaved deciduous woodland there are small areas of heathland, grassland, bog and marsh. The managed high-forest and areas of well-established wood-pasture are utilised by a small population of **Barbastelles**.

15 Strumble Head and the Pembrokeshire coast

Grid ref: SM897413 (Strumble Head)

Key species

Harbour Porpoise, Risso's Dolphin, Killer Whale, Grey Seal.

Access

Strumble Head can be reached via minor roads from Fishguard.

Harbour Porpoise: Systematic recording of cetaceans off Strumble Head began in 1992. Since then, Harbour Porpoises have been shown to peak in numbers during April with numbers reaching at least 50. Numbers drop off in May, June and July, before increasing rapidly to another peak in August and September. In August up to 700 porpoises have been in view at one time. During these periods, days when no porpoises are seen are very rare. In winter, they prove more difficult to locate with the rougher weather conditions and the absence of Northern Gannets, which wheel conspicuously over the feeding groups of porpoises during the summer months.

Risso's Dolphin: Risso's Dolphins are regularly reported from headlands on the Pembrokeshire mainland as well as from the offshore islands of Skokholm, Skomer and Ramsey. Most sightings occur between May and October, although in recent times there have been reports of groups passing Strumble Head during winter. Schools normally number between two and 12 individuals.

Killer Whale: Killer Whales are recorded infrequently off the Pembrokeshire coast. When they do occur, it is often during September coinciding with the onset of the Grey Seal pupping season. It may be worthwhile watching areas of coast that contain Grey Seal colonies around the time of pupping, as any may attract Killer Whales looking for easy prey. They have also been observed off prominent headlands such as Strumble Head during the spring, usually in March, as they head up the Irish Sea into more northerly waters to spend the summer.

Grey Seal: The largest Grey Seal colony in south-west Britain, approximately 5,000 individuals, breeds around the rocky coast and islands of Pembrokeshire. They can be viewed from most points of the Pembrokeshire coast, as well as from the islands of Skokholm, Ramsey and Skomer.

16 Stackpole Estate, Pembrokeshire

Grid ref: SR977962
(Greater Horseshoe Bat roost)

Key species

Otter, Greater Horseshoe Bat, Whiskered Bat, Daubenton's Bat.

Access

This site is signposted off the B4319, five miles (8km) south of Pembroke. From Stackpole village head towards Bosherston and you will drop down into a wooded valley. At the bottom of the valley turn left over a small bridge, drive a short way up the hill and park in the obvious parking spot.

Site description

To the right of the parking spot is a large lawn beyond which is an old stable block with a high archway in the middle. The bats emerge from the broken ceiling of the archway. The people who live in the stable block all seem to know about the bats, but it is still best to ask at one of the houses for permission to stand/sit on the lawn to watch the emergence. Stand well back to avoid any disturbance to the colony. The bats can be identified without a bat detector but take one along (if you have one) as the echolocation sounds are fantastic.

Daubenton's Bat (*Dan Brown*)

This accessible roost site holds approximately 9.5% of the UK population of Greater Horseshoe Bats. It represents the species at the northwestern extremity of its range.

In addition to the Greater Horseshoes several other species of bats occur here. The bridge mentioned above can be excellent for **Daubenton's Bats** and **pipistrelles**. **Whiskered Bats** also occur.

Otter: Bosherston Ponds offer an opportunity for viewing Otters. In the past the lily-ponds were partially estuarine before damming created the ponds that can be seen today. Search out the quieter creeks at dusk to increase the chance of success. There is a hide on the western side of the ponds (SR977058). There is ample car parking space.

17 Pencelly/Pengelli Forest, Pembrokeshire

Grid ref: SN130390

Key species

Common Pipistrelle, Soprano Pipistrelle, Brown Long-eared Bat, Barbastelle, Natterer's Bat.

Access

The forest is west of Cardigan and just north of the A487. Between the B4329 turn off and Newport, take the minor road immediately east of Felindre Farchog. There is very limited parking on the roadside at the entrance to the wood at SN123395. There is a path into the wood on the east side of the road.

Site description

Ancient woodland on a steep hillside. The wood is an SSSI and is a West Wales Wildlife Trust reserve. Many of the paths are steep, muddy and difficult to walk but the site is good for a range of bat species. In addition to those listed above, **Greater Horseshoe Bats** are also occasionally recorded.

18 Dyfi, Gwynedd

Grid ref: SH750046

Key species

Hazel Dormouse, Fallow Deer.

Access

The area can be accessed north of Machynlleth via the A487 that bisects the heart of this large forest.

Site description

Lying to the east of Cader Idris, large areas of Dyfi Forest were deforested during the peak in slate mining activity in the area. Such extensive activity has left many disused mine workings in the area and these now provide amongst other things, roosts for bats and niches for uncommon plant species. The Forestry Commission now owns and manages Dyfi Forest, and they have reforested much of the area post-mining. **Hazel Dormice** occur within the forest, although they are difficult to spot.

Fallow Deer can be found among the upland oaks that predominate here. **Roe Deer** also occur.

19 Tan y Coed, Gwynedd

Grid ref: SH755053

Key species

Badger.

Access

This site is four miles (6km) north of Machynlleth and can be accessed west off the A487 south of Corris and north of Pantperthog. There is a pay-and-display car park, toilets and a picnic area. There are several way-marked trails as well as bridleways and forest roads accessible to cyclists and horses.

Site description

This picnic site is located amongst beautiful Forestry Commission broadleaved woodland. There are several **Badger** setts within the woods, and the animals can often be observed foraging on warm summer evenings.

20 Coed y Brenin, Gwynedd

Grid ref: SH716277
(Forestry Commission visitor centre)

Key species

Lesser Horseshoe Bat, Whiskered Bat, other common bat species, Red Squirrel, Fallow Deer.

Access

The A470 between Trawsfynydd and Llanelltyd runs through the heart of the forest park. The Forestry Commission operates a visitors' centre at Maesgwm where there is a pay-and-display car park and toilets. The centre is open from 10:00 to 17:00 hrs from Easter until October and at weekends during the rest of the year. There are various car parks throughout the forest, as well as further toilets.

Site description

This diverse area contains a mix of habitats and species, mainly due to the varied geology, topography and mild mid-Welsh climate.

Bats: This area of mixed woodland provides habitat for many species including **Common** and **Soprano Pipistrelles**, **Noctules**, **Whiskered Bats**, **Brown Long-eared Bats** and **Lesser Horseshoe Bats**. The best places to view bats are over the rivers that run through the forest, particularly the Mawddach. Guided bat walks are organised by local bat groups during the summer and it is best to join one of these in order to locate the best areas within the forest for every species, as well as for help in identifying them. See 'Bat Groups' for contact details.

Red Squirrel: This species occurs but is present at an extremely low density. The Forestry Commission Conservation Manager would like to receive any reports, so please pass on all sightings to the Coed y Mynydd office, tel: 01341 422289.

Fallow Deer: Both **Fallow Deer** and **Roe Deer** occur in the forest, but the former are undoubtedly commoner and more frequently observed. The southern sections of the forest usually produce the best results.

There have been unconfirmed reports of **Pine Martens** for a number of years and a project has been initiated to try and confirm their presence.

21 Penmaenuchaf Hall Hotel, Penmaenpool, Gwynedd

Grid ref: SH697184

Key species

Lesser Horseshoe Bat.

Access

Penmaenuchaf Hall Hotel is located in the village of Penmaenpool, approximately 3 miles west of Dolgellau. To watch the bats it is necessary to stay at the hotel. Further details can be found at the hotel's website: www.penhall.co.uk.

Site description

This is an excellent place to watch large numbers of **Lesser Horseshoe Bats**. The best time to visit is at the end of July and beginning of August as the roost can hold up to 250 bats at this time, when juveniles join their mothers on the wing. They can be watched at close range through a special observation window that overlooks the entrance to the roost. It is also possible to listen to them in the hotel garden as they fly over the kitchen roof and into the surrounding woodland. The inexperienced youngsters often cling onto external walls and can provide exceptional views, particularly through binoculars.

22 Snowdonia, Gwynedd

Key species

Feral Goat.

Site description

A survey in 1991 revealed a total of 282 Feral Goats in Snowdonia. A good place to see them is Cwm Idwal, about five miles (8km) west of Capel Curig by the A5 (SH723580): park west of the A5 just north of the A4086 junction. Another good site is the old slate quarry above Padarn Country Park, Llanberis (SH584608).

23 The Lleyn Peninsula and Bardsey Island, Gwynedd

Grid ref: SH118218 (Bardsey Island)
National Trust parking facilities on Lleyn Peninsula tip: SH139259

Key species

Risso's Dolphin, Harbour Porpoise, Grey Seal.

Access

The Lleyn Peninsula is easily accessed by road from points eastward via Porthmadog or Caernarfon. Bardsey Island can be accessed by boat during the summer only. Trips usually start out from Porth Meudwy, but may be switched to from Pwllheli. Rough seas are common in Bardsey Sound and often force the cancellation of trips, particularly late in the season, so have a back-up plan.

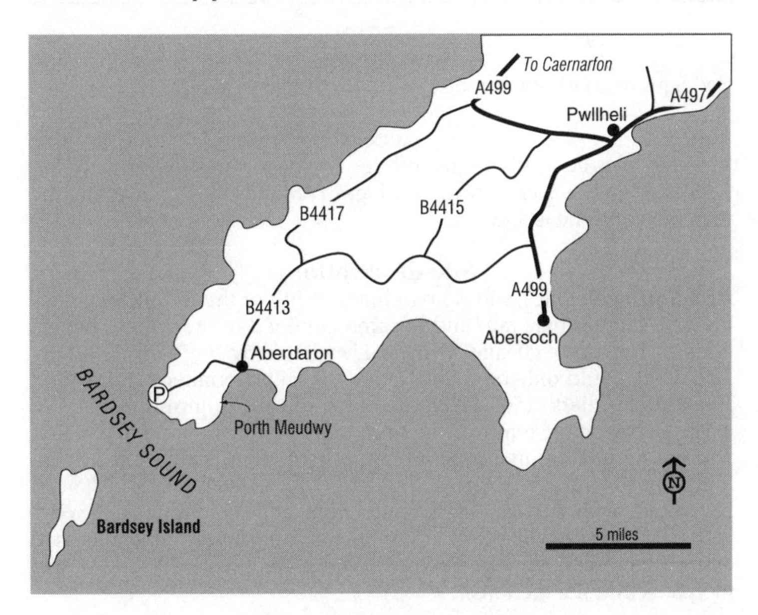

Site description

The south-western-pointing Lleyn Peninsula and Bardsey Island, located off its tip, produce almost as many **Risso's Dolphin** sightings as Pembrokeshire. Like that county, the peak times are between May and October, when small schools can be seen passing any headland in the west of the peninsula. **Harbour Porpoises** are still relatively common in the seas off the Lleyn Peninsula and around Bardsey Island, and with calm seas at the right time of year you should be able to enjoy sightings of our smallest cetacean species.

A small number of **Grey Seals** breed within rocky bays on Bardsey each year and these are usually readily viewable on a walk around the island.

24 Coastal Gwynedd

Key species

Polecat.

Site description

The lowland areas of coastal Gwynedd support large populations of **Polecats**. Areas between Harlech and Barmouth and Caernarfon and Bangor are particularly worth trying at dusk and during the night when driving slowly down minor roads in either area can produce sightings.

25 Anglesey

Key species

Red Squirrel, Harbour Porpoise.

Access

Anglesey is a large island just off the Gwynedd coast in north Wales. Access is simple, with two road bridges (A5 and A55) connecting the mainland and the island.

Site description

Red Squirrel: Once widespread in the south of the island, they have become increasingly rare and became confined to the conifer plantations of Newborough and Mynydd Llwydiarth by the early 1990s. By 1997 Reds could only be found within the 240ha conifer plantation of Mynydd Llwydiarth (SH544788). Trapping of Grey Squirrels resulted in 4,500 Greys being removed in four years, following which the Red Squirrel population increased from a low of 40 individuals in 1998 to between 80 and 95 in the spring of 2002. Most occur in Mynydd Llwydiarth with occasional sightings from adjacent broadleaf woodlands. Mynydd Llwydiarth is in the east of the island and is accessed east off the A5025 through Pentraeth. There are forest trails and public footpaths through the forest.

26 Moel Famau, Denbighshire

Grid ref: SJ161627

Key species

Polecat.

Access

From Mold take the A494 through the towns of Gwernymynydd and Loggerheads until you come to a junction with 'Moel Famau Country Park' signposted to the right. Follow this road for a mile (1.6km) to a car park on the right-hand side.

Site description

The Forestry Commission woodland of Moel Famau is on the Clwydian Range, and has been described as the gateway to rural Wales. The woodland covers some 579 hectares. The **Polecat** is best looked for along the woodland edges and in woodland clearings. Walking the tracks at night may also prove fruitful. As well as the target species, other mammals present at Moel Famau include **Stoats**, **Weasels**, **Brown Hares, Badgers** and **Red Foxes**.

27 Nercwys, Flintshire

Grid ref: SJ218593

Key species

Muntjac, Polecat, Red Fox.

Access

From Mold go into Nerwcys village and head straight through until you come to a crossroads. Turn right here and you will come along the forest edge. Take the first right and continue for approximately 500m to the car park. The woodland is relatively flat, and the extensive network of footpaths and rides allows access for walkers, cyclists or horse riders.

Site description

This Forestry Commission managed woodland was mainly planted during the 1950s and 1960s and now consists of an extensive area of coniferous trees. Small populations of **Muntjac**, **Polecats** and **Red Foxes** may be found within the woodland. They are best seen by walking quietly along the trails at dusk.

SCOTLAND

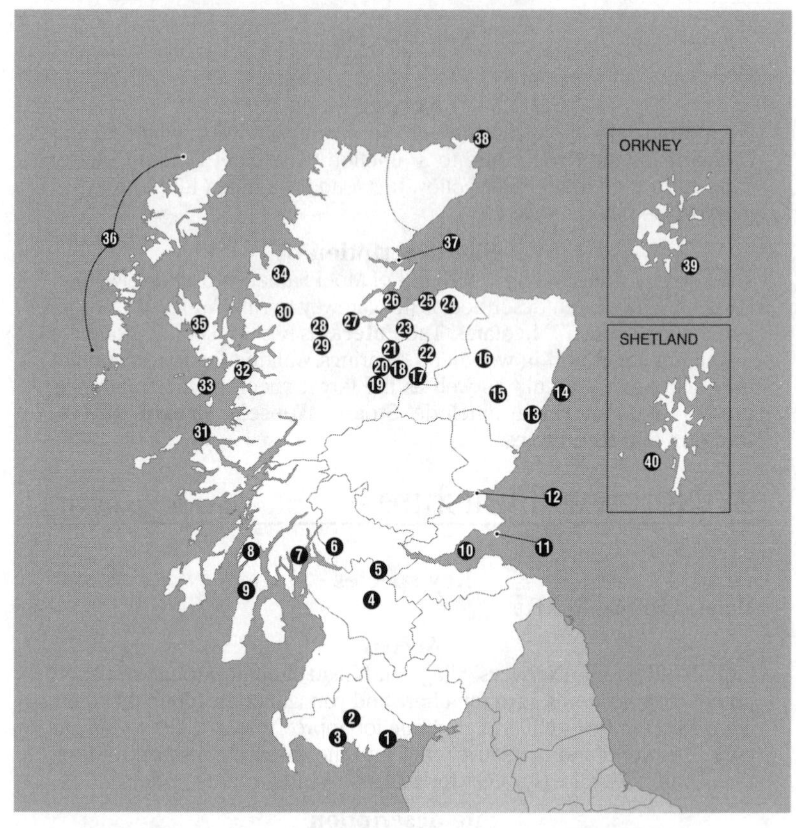

1 Mabie Forest
2 Cairnsmore of Fleet
3 Penninghame Pond
4 Falls of Clyde
5 Palacerigg Country Park
6 Loch Lomond and Trossachs National Park
7 Kilmun Arboretum
8 Ardcastle
9 Knapdale
10 Firth of Forth
11 Isle of May
12 Firth of Tay and Eden Estuary
13 Durris Forest
14 Aberdeenshire coast
15 Glen Tanar
16 Clashindarroch
17 The Cairngorms
18 Ryvoan Pass, near Aviemore
19 Insh Marshes RSPB Reserve
20 Abernethy Forest RSPB/NNR
21 Slochd Summit
22 Lynemore
23 Findhorn Valley
24 Monaughty
25 Culbin
26 Moray Firth and Cromarty Firth
27 Glen Urquhart
28 Glen Affric
29 Plodda
30 Achnashellach
31 Beinn Eighe NNR
32 Kinloch
33 Inner Hebrides and mainland Argyll and Bute
34 Ardnamurchan Peninsula
35 Isle of Skye
36 Outer Hebrides
37 Dornoch Firth
38 Duncansby Head
39 Orkney
40 Shetland

1 Mabie Forest, Dumfries and Galloway

Grid ref: NX949708 (Mabie Forest car park)

Key species

Red Squirrel, Roe Deer.

Access

This 'Community Forest' overlooking the Solway Firth caters for visitors of all abilities with wheelchair accessible paths along many of the longer trails. Mabie Forest lies four miles (6km) from Dumfries on the A710. Follow the signs for the Solway coast. From the village of Islesteps travel a further two miles (3km) and the forest entrance is on the right-hand side at the corner of a tight bend.

Site description

Mabie Forest is home to a wide variety of mammals. **Red Squirrels** can often be seen close to the car park and near to the squirrel feeders on the All Abilities Walk. **Roe Deer** can also sometimes be seen in the forest, particularly on the longer trails early in the morning. There are five species of bat present in Mabie Forest, with **pipistrelles** being the most frequently recorded. Other mammals present include **Red Foxes**, **Pine Martens**, **Water Voles, Bank Voles, Wood Mice, Common Shrews** and **Hedgehogs**.

2 Cairnsmore of Fleet, Dumfries and Galloway

Grid ref: NX491721

Key species

Feral Goat.

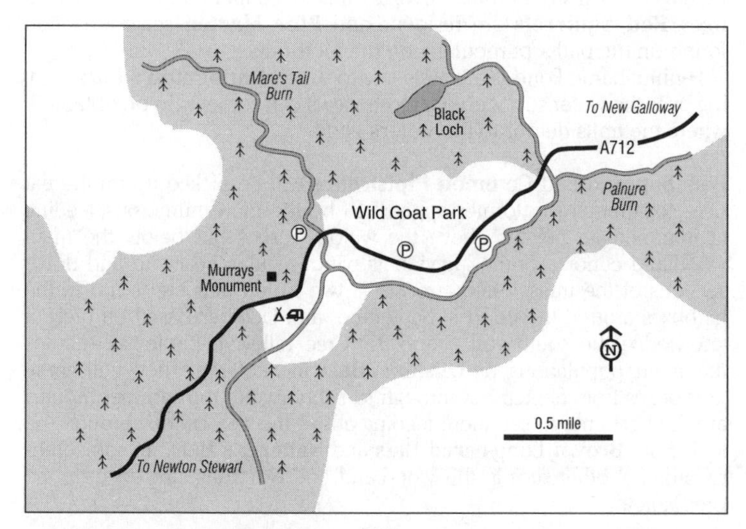

Access
Cairnsmore of Fleet is on the A712 between Newton Stewart and New Galloway.

Site description
This large expanse of undulating moorland holds a herd of easily viewable **Feral Goats**. Park in the car park at the above grid reference and scan the cliffs above the road. Another herd is present nearby in the Galloway Forest Park, near Newton Stewart.

3 Penninghame Pond, Newton Stewart, Dumfries and Galloway **Grid ref: NX375692**

Key species
Daubenton's Bat, Leisler's Bat, Soprano Pipistrelle, Common Pipistrelle.

Access
From Newton Stewart, take the A714 to Girvan and after 2½ miles take a left turn (signposted 'Penninghame Pond'), onto a Forestry Commission road. Follow this road for a further 500m to a small car park beside the pond. The forest road and car park are open at all times. An all-abilities trail leads from the car park, around the pond.

Site description
The pond has resulted from the damming of a natural watercourse, and was originally developed to provide a source of hydropower for the estate sawmill. At the downstream exit, a fish pass has been constructed to enable Atlantic Salmon to continue to access their spawning grounds upstream. **Otters** may sometimes be seen along this stream. The area is an ancient woodland site, now owned by the Forestry Commission and managed for a continuous cover of mixed conifers and broadleaved trees. **Red Squirrels** are resident, and **Pine Marten** scats are often found on the paths, particularly on the footbridge.

Penninghame Pond is a classic site to watch **Daubenton's Bats** feeding over the water surface: very close views can sometimes be obtained, where the trails detour to the water's edge.

Both **Soprano** and **Common Pipistrelles** will be picked up on the bat detector, though the former appears to be the more numerous, feeding at intermediate heights above the willow bushes but below the main woodland canopy. Some individuals patrol regular 'beats' up and down sections of the trails. Mating roosts of two to ten bats are found in the batboxes around the pond in September and October. Based on present knowledge, the south-west – and the Cree Valley in particular – holds the main populations of **Leisler's Bats** in Scotland. They utilise the nest boxes from September through to February, but the summer months are the best time to see them feeding above the tree canopy around the loch. Both **Brown Long-eared Bats** and **Natterer's Bats** are sometimes recorded at other sites in this woodland, and **Noctules** also occur in the Cree Valley.

4 Falls of Clyde, Clyde **Grid ref: NS882425**

Key species

Natterer's Bat.

Access

The Falls of Clyde Nature Reserve lies approximately one mile (1.6km) south of Lanark and is accessed via New Lanark, which is signposted from all major routes. From Glasgow, take either the A72 or A744; from Edinburgh, take the A743; from Ayrshire and the south, take junction 12 of the M74. The reserve has a visitors' centre that is open from 12:00 to 16:00 hrs in January and February and from 11:00 to 17:00 hrs from March to December. The reserve itself is open from 08:00 to 20:00 hrs in summer and during daylight hours in winter.

Site description

This 59-hectare Scottish Wildlife Trust reserve is home to some beautiful scenery. Away from the waterfall, there is some good woodland along both sides of the River Clyde where **Natterer's Bats** may be seen hunting. **Red Foxes**, **Badgers** and **Roe Deer** also occur at this site.

5 Palacerigg Country Park, Clyde **Grid ref: NS785733**

Key species

Brown Hare, Red Fox, Roe Deer.

Access

Palacerigg Country Park is on the southern outskirts of Cumbernauld and is signposted off the A8011 Cumbernauld Town Centre road. The visitors' centre is open all year (April to September: 10:00–18:00 hrs, October to March: 10:00–16:00 hrs). Both the centre and toilets have disabled access. There is a good network of paths around the park.

Site description

This diverse park consists of moorland, farms and areas of woodland. It was established in the 1970s by Cumbernauld Borough Council. Over 40 hectares of what was once an upland farm have been planted with thousands of native trees and shrubs. **Brown Hares** are best looked for on the areas of grassland, and they are usually easy to see, particularly early in the morning or late in the evening when most visitors have left the park. **Roe Deer** occur throughout the areas of farmland and woodland, whilst **Red Foxes**, **Badgers** and **Rabbits** also occur here. Small populations of **Red Deer, Sika Deer, Fallow Deer** and **Muntjac** are maintained in the pasture/parkland areas.

6 Loch Lomond and Trossachs National Park, Clyde

Grid ref: NS378889

Key species

Red-necked Wallaby, Brown Long-eared Bat.

Access

The national park is just an hour's drive north from Glasgow via the A82.

Site description

The large national park is home to a diverse range of habitats and species. Loch Lomond, Britain's largest expanse of fresh water, dominates the area, but the park also encompasses mountains, glens, farmland and large tracts of woodland. **Brown Long-eared Bats** are thinly distributed throughout where suitable habitat is available. Other species present include large herds of **Red Deer**, **Red** and **Grey Squirrels**, **Water Voles** and **Water Shrews**.

Red-necked Wallaby: Red-necked Wallabies were first introduced to Loch Lomond in the 1970s. Lady Arran maintained a colony on Inchconnachan, one of the larger islands on the loch, for a number of years with the intention of turning it into a wildlife park. However, the wallabies were soon left to their own devices. They adapted to the Scottish climate and seemingly flourished. In severe winters, some would hop across the frozen loch surface to live in the woodlands at the loch-side. A report by Scottish Natural Heritage in 1998 stated that there were 28 Red-necked Wallabies on Inchconnachan. Later reports in local newspapers indicated some individuals were seeking pastures new and were swimming across to the mainland. The majority of the sightings were centred on the village of Luss, on the western shore of the loch. It appeared the island, which is only about three-quarters of a mile (1km) long and 500 yards (450m) wide, was getting a bit cramped for the expanding colony of around 40 wallabies and that they were looking to colonise new areas. On the mainland, they appeared very shy and soon disappeared when alarmed: there are none there at present.

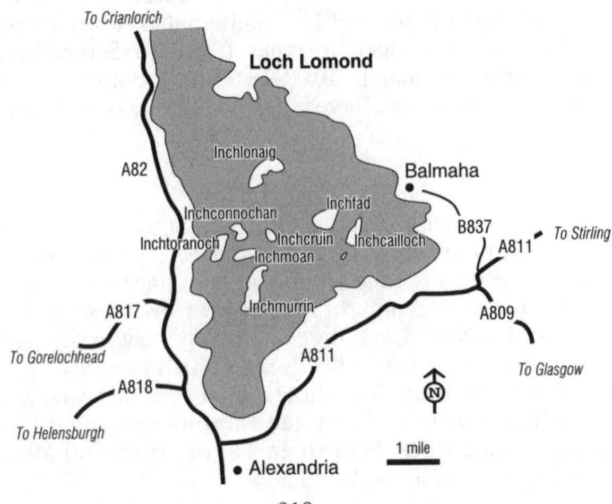

To see the animals now, it is necessary to get to Inchconnachan. There are two ways of getting to the island, either by hiring a boat yourself, or by arranging with a local tour operator in Luss to drop you off on one of the morning cruises, and to pick you up again at a pre-agreed time later in the day. Private boats can be hired from the local boatyard on the eastern side of the loch in the village of Balmaha. They cost £40 for the day and seat a maximum of five persons. Make sure you have a map, as the well-vegetated islands, many of which are not included on large-scale maps, tend to blend into the background and can have you guessing at what's mainland and what's not.

Inchconnachan has a small jetty and beach suitable for landing. During the summer, the southern side of the island tends to have a lot of campers and is best avoided. Head right immediately from the beach into the best areas. The runs and faeces of the wallabies are obvious. There are only a couple of tracks on this side of the island and a wrong turn may have you battling through the undergrowth. Although some people have taken some time to find their first wallaby, others have succeeded almost immediately and have recorded eight to ten individuals in under an hour. Walking slowly down the trails will normally provide good views. Further information may be obtained from the Loch Lomond Ranger, C.Calvey (01389 753311).

7 Kilmun Arboretum, Argyll and Bute

Grid ref: NS164823 (Kilmun car park)

Key species

Red Deer, Red Squirrel, Roe Deer.

Access

Kilmun Arboretum is about 25 miles (40km) west of Glasgow. It can be reached from Glasgow by car ferry from Gourock to Dunoon, or via the A82 along Loch Lomond through Tarbet, towards Arrochar. From here, follow signs for Kilmun. The arboretum is generally well signposted from all directions.

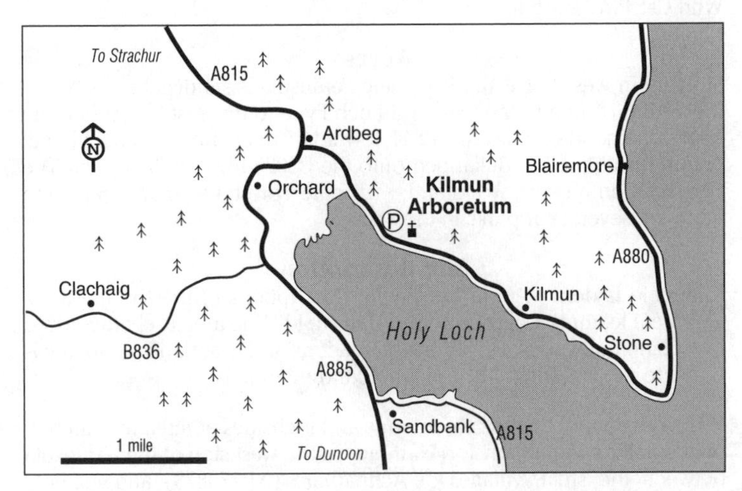

Site description

Initially established to monitor the success or otherwise of exotic tree species on the west coast of Scotland, Kilmun Arboretum now offers refuge for good numbers of **Red Deer**. The coniferous forests offer ideal conditions for the deer and from time to time they become so numerous that the Forestry Commission has to cull them humanely. Other mammals that may be seen in the woodlands include **Red Squirrel** and **Roe Deer**, both of which are common.

8 Ardcastle, Argyll and Bute

Grid ref: NR941923
(Ardcastle car park)

Key species

Pine Marten, Red Squirrel.

Access

Ardcastle Woods are just off the A83 between Minard and Loch Gair. Several trails start at Ardcastle car park.

Site description

This site came under ownership of the Forestry Commission in 1938 and much of the area has been extensively planted with conifers. The woodlands are home to many typical Scottish woodland species. As well as the target species listed above, you may also come across **Roe Deer**, **Red Deer** and **Red Foxes**.

9 Knapdale, Argyll and Bute

Grid ref: NR714640
(Kilberry, Knapdale Peninsula)

Key species

Wild Cat, Pine Marten.

Access

Situated in west Argyll, the Knapdale Peninsula is sandwiched between the Sound of Jura to the west and Loch Fyne to the east. To get into the heart of Knapdale take the B841 towards Crinan from Lochgilphead, before turning left at Bellanoch onto the B8025 towards Tayvallich. The Forestry Commission woodland, Knapdale Forest, has many trails and there are several car parks in the area.

Site description

Knapdale is derived from the Gaelic descriptions of the striking topographical features: Cnap (hill) and Dall (field). The area is characterised by upland oakwoods, coniferous plantations and moorland. Knapdale is home to the majority of British woodland mammals.

Wild Cats patrol the oak woodlands and lochsides at night in search of their small mammal prey, for example, on the west side of the peninsula between the small villages of Achnamara (NR779873) and Ashfield

(NR765854) around the meadows that abut the sea-loch where **Rabbits** abound. **Common Seals** may also be seen on the sea-loch from the road here. Another good site for **Wild Cats** is Kilmichael Forest. This is a huge working forest covering 75km^2, and away from the commercial operations much of it provides sanctuary for many mammal species. **Pine Martens**, **Badgers, Red Foxes, Red Deer, Roe Deer** and **Otters** may also be found as well as Wild Cats. The forest is sandwiched between the villages of Lochgilphead and Minard, with the A83 constituting the south-eastern boundary. The best access points are at Achnabreck – three miles (nearly 5km) north of Lochgilphead turn right onto the unsurfaced forest road for 500 yards before another right turn into the car park – and Glashan: from Minard follow the A83 for seven miles (11km) to where there is a car park.

10 Firth of Forth, Forth

Grid ref: NT128795 (Forth Road Bridge)

Key species

Harbour Porpoise.

Access

This large area is easily watched from either the north or south shores. Good spots include North Queensferry (north shore) and Queensferry westwards (south shore).

Site description

In 2003 a single **Humpback Whale** spent two weeks in the waters around the Forth Road Bridge. According to local observers this was the third consecutive year in which the species has appeared in March apparently following shoals of Herring into the Firth of Forth. **Harbour Porpoises** are regularly seen in the same area.

11 Isle of May, Forth

Grid ref: NT654992

Key species

Grey Seal.

Access

Visitors are welcome to visit the island between April and September, with the *May Princess* leaving from the mainland port of Anstruther between May and September. The authorities ask that visitors keep to the paths to reduce disturbance and to keep a respectable distance from the cliff edges, both for their own safety and to reduce the possibility of disturbing the cliff-nesting seabirds. There are a number of well-marked paths as well as an information centre.

Site description

Lying at the mouth of the Firth of Forth, the Isle of May is the largest east coast breeding colony of Grey Seals in Scotland, and the fourth largest breeding colony in the UK with approximately 1,300 pups being born here in 1996. At the start of 2001, the total population of Grey Seals on the island numbered 7,000.

12 Firth of Tay and Eden Estuary, Angus/Fife, Perth and Kinross

Grid ref: NO498192 (Out Head car park), Eden Estuary). Tayport: NO460287

Key species

Common Seal.

Access

The seals may be watched from various points at both locations but good spots include the Out Head car park, on the southern shore of the Eden Estuary, and from Tayport westwards, on the southern shore of the Firth of Tay.

Site description

The Firth of Tay and Eden Estuary support a nationally important breeding colony of Common Seals. Around 600 individuals, some 2% of the UK population, haul-out on sand banks at the site.

13 Durris Forest, Aberdeenshire

Grid ref: NO786926

Key species

Red Squirrel, Roe Deer.

Access

Durris Forest lies south of the River Dee between Banchory and Maryculter. It can be accessed from Stonehaven or Banchory east off the A957. There are various points of entry to the woodlands:

- Inchloan Entrance: NO769924
- Curly Brae Entrance: NO783948, which has room for five or six cars
- Slug Entrance: NO781908, has similar capacity to Curly Brae Entrance and has many paths that are popular with walkers
- The Café Entrance: NO773918. Footpaths are plentiful throughout the woodlands and there is also a way-marked cycle trail.

Site description

The Durris Forest encompasses the forest itself and a number of outlying smaller woods. **Roe Deer** are regularly seen within the forest, and your chances are increased with regular visits, particularly at dawn and dusk. **Red Squirrels** thrive in the conifers of Durris Forest. Much of the present management work, which consists of age-structuring the forest, should benefit this species. **Red Foxes** and **Red Deer** occur throughout the forest.

14 Aberdeenshire coast

Key species

White-beaked Dolphin, Bottle-nosed Dolphin, Minke Whale, Harbour Porpoise.

Access

The Aberdeenshire coastline offers opportunities to see a range of species. An excellent detailed article on cetacean watching in this area was published in *Whale and Dolphin* magazine in May 2005 (Weir, 2005) and is the essential reference for anyone visiting this area. Good vantage points along the coast include:

- the fishing port of Stonehaven 15 miles (24km) to the south of Aberdeen
- the coastal footpath south of Stonehaven
- Girdleness Lighthouse on the headland south of Aberdeen Harbour
- the coastal path between Cove Bay and Aberdeen City
- Nigg Bay
- Souter Head north of Cove
- Greg Ness.

Boat trips based in Stonehaven Harbour run regular summer trips along the coast to Fowlsheugh RSPB Reserve and frequently produce sightings of White-beaked and Bottle-nosed Dolphins. Contact either Brian Bartlett on 01569 766040, or Ian Watson on 01569 765064 for details.

Site description

The summer months off the coast of east Scotland can provide the opportunity for some exciting cetacean spotting. **Harbour Porpoises** can be seen all year but particularly in June to September, **Bottle-nosed Dolphins**, December to May, **White-beaked Dolphins**, June to September, and **Minke Whales** are regularly seen from June to August.

Killer Whales, Common Dolphins, Atlantic White-sided Dolphins and **Long-finned Pilot Whales** are seen infrequently and recent rarities have included **Humpback, Fin** and **Sperm Whales**.

Other information

Details of recent cetacean sightings are available on www.wildlifeweb .force9.co.uk and sightings should be reported to the South Grampian Sea Watch Group at southgrampianseawatch@hotmail.com.

15 Glen Tanar, Highland

Grid ref: NO459927

Key species

Wild Cat.

Access

From Stonehaven, to the south of Aberdeen, follow signs for the A957 to Banchory. At Crathes turn left onto the A93 through Banchory towards Aboyne. In Aboyne turn left at the village green, left at a T-junction and then continue over the River Dee. Follow signs to Glen Tanar along the B976 and turn left at Tower of Ess into the estate.

Site description

This nature reserve, managed by Scottish Natural Heritage, contains a significant tract of ancient Caledonian pine forest. **Wild Cats** are the top predators on the estate. As well as the areas of pine forest, the estate also consists of sections of broadleaved woodland and moorland higher up on the hills. In the 1970s the Wild Cat population on the estate was estimated at 30 animals per 100 square kilometres, much higher than for populations in western Scotland, although this seems likely to have been an overestimate.

There is a visitors' centre at Braeloine with a car park and other facilities. A network of trails radiate from the visitor centre and these provide suitable walks for all abilities. **Red Deer** also occur here on the moorland, and many other common Scottish mammals may be found on the estate.

Other information

Parrot Crossbills and a small population of Capercaillies may also be found in the woodlands.

16 Clashindarroch, Aberdeenshire

Grid ref: NJ429268

Key species

Red Deer, Red Squirrel.

Access

To access the area follow the A941 from Rhynie to Dufftown for approximately five miles (8km). There is a large car parking area on the left.

Site description

These forests are the best snow-holding woods in the country, and there are several specially laid-out trails for cross-country skiing. Herds of **Red Deer** are often seen in winter around the Cabrach car park at Clashindarroch. **Red Squirrels** and **Roe Deer** are also found in the area.

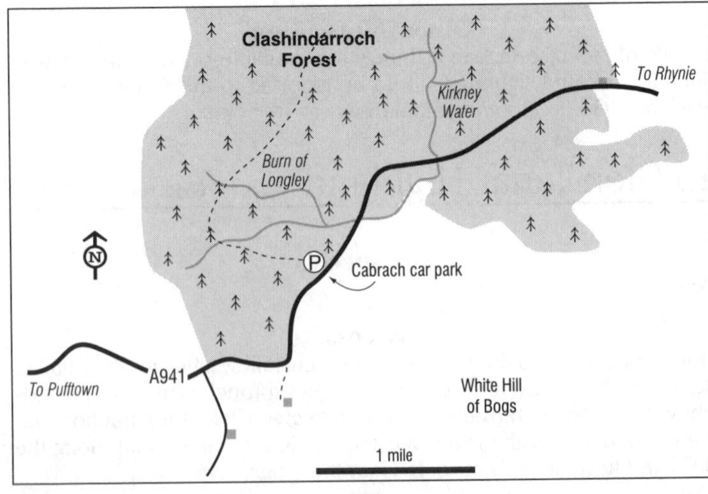

17 The Cairngorms

Grid ref: NH986054 (Cairngorm Chairlift and Ski Centre)

Key species

Caribou (Reindeer), Mountain Hare.

Access

The Cairngorms are well signposted from the A9 and are accessed on minor roads through Glenmore. The plateau can be accessed via the new funicular railway or ski chairlift.

Site description

Reindeer: The Reindeer that now reside on the Cairngorm plateau can be viewed by joining a guided walk with the 'Reindeer Company', or, if time is limited, they can easily be seen from the road that meanders up to the Cairngorm ski station, accessed east off the B970 through Glenmore. From the mountain road look west just above the tree line and small numbers are usually visible within their winter enclosure. More aesthetically pleasing free-ranging individuals can sometimes be seen by walking out across the plateau from the uppermost railway or chairlift station.

Mountain Hare: The Cairngorm plateau provides good Mountain Hare habitat, and they can usually be easily observed anywhere on the plateau, although good spots include the area between Cairngorm summit and Bynack Moor, or from the trail to Ben Macdui (NN989989).

18 Ryvoan Pass, near Aviemore, Highland

Grid ref: Starting point (on Glen More road): NH982087. Ryvoan Bothy (mid-point of trail): NJ005115. Forest Lodge, Abernethy Forest (finishing point): NJ019160.

Key species

Wild Cat.

Access

Ryvoan Pass can be reached from Glen More at the foot of the Cairngorms, and the track through the pass meanders over some low hills into Abernethy Forest.

Site description

Good areas to try are where the moorland and woodland meet. There are grassy areas as well that may attract **Wild Cats** due to higher prey densities.

19 Insh Marshes RSPB Reserve, Highland

Grid ref: NH799020

Key species

Otter.

Access

Insh Marshes are located two miles (3km) east of Kingussie on the B970. There are three trails around the reserve but unfortunately due to the rocky terrain none of them are passable with a wheelchair. Good views of the reserve can also be obtained from the B970. Other facilities on site include two hides and an information viewpoint in the car park.

Site description

The Insh Marshes cover 844 hectares and are one of the most important wetlands in Europe. The extensive marshes together with lochs, small lochans and ditches provide ideal feeding, resting and shelter areas for **Otters**, and the area supports a healthy population.

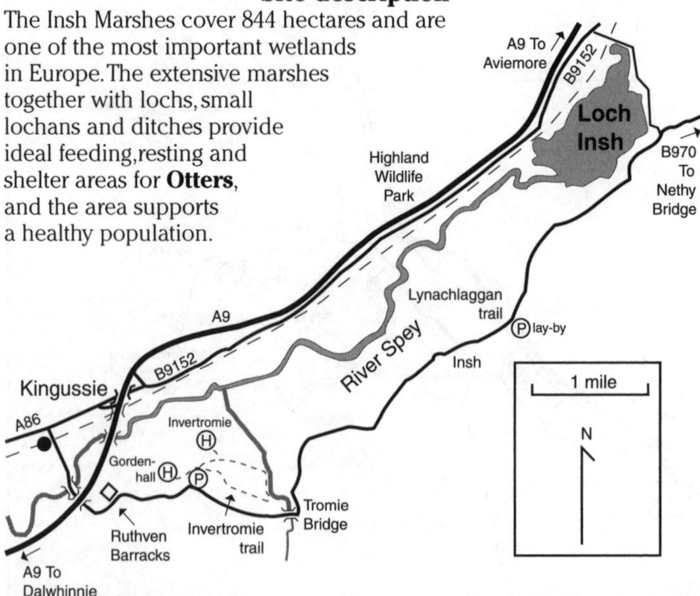

20 Abernethy Forest RSPB/NNR, Highland

Grid ref: NH977184

Key species

Red Squirrel, Wild Cat, Red Deer, Pine Marten, Badger.

Access

From Aviemore take the A95 north before following RSPB signs (Osprey signs in use in spring and summer only) through Boat of Garten to the reserve. The grid reference is for the Osprey centre car park although there are numerous pull-ins throughout the forest as well as trails for the disabled and other footpaths.

Site description

A visit to the Abernethy Forest visitors' centre at Loch Garten is a must for anyone wishing to experience some of Scotland's most special wildlife. It is the largest native Scot's Pine forest left in Britain, and the mix of woodland and northern bog combine to offer sanctuary for many specialist creatures.

Red Squirrel: This species is relatively easy to see in the forest. They can occur throughout the area but are frequently seen from the path to the visitors' centre and from the road around the edges of Loch Garten. Nearby, the Carrbridge Landmark Centre, just north of Aviemore, is another good spot for this species and they are easily seen in the forest around the visitors' centre there.

Red Deer: Red Deer are extremely common within the pinewoods and heathland of Abernethy Forest. Indeed, their population has grown to such an extent that their numbers are seriously reducing natural regeneration of the woods, which in turn is having detrimental effects on the whole plant and animal community.

Wild Cat: Although typically difficult to see, Wild Cats do occur within the forest although the chances of encountering them are slim. Driving the roads at night offers the best chance of seeing one.

Wild Cat (*F Müller*)

Pine Marten: The best way to track this species down within Abernethy is from the RSPB Pine Marten Hide, particularly during June. Entrance to the hide, however, is fairly expensive (see the RSPB website for details). They may also be baited to well-used car parks within the forest. Another ploy is to drive slowly at night along the forest roads looking out for eye shine in the trees or undergrowth. If you are lucky, one may run across the road in front of you!

Badger: The Strathspey Badger Hide, three miles (nearly 5km) from Boat of Garten, was established in 1996 to provide nature enthusiasts with the chance to observe the species at close quarters. There is a nominal charge for visiting but the success rates are usually quite high due to the presence of a Badger family within the area. It is open from the middle of March to the end of summer. Visits are possible by contacting the hide manager Allan Bantick by email: bantick@aol.com.

Other species in the area include **Weasels**, **Stoats**, **Red Foxes**, **Daubenton's Bats** and **Brown Long-eared Bats**.

Other information

During the last few springs a Capercaillie lek has been viewable from the visitors' centre, and those who wish to see this nationally rare species are encouraged to attend these watches in order to reduce disturbance to other leks within the forest.

21 Slochd Summit, Highland **Grid ref: NH801292**

Key species

Mountain Hare.

Access

Slochd Summit is north of Aviemore and straddles the A9. To access the best areas leave the A9 and head west towards Tomatin. Immediately after crossing the railway, park in the long lay-by to the south of the road.

Site description

From the lay-by scan the hillsides on the opposite side of the road. **Mountain Hares** can be seen here throughout the year, with large numbers present in spring.

22 Lynemore, Highland **Grid ref: NJ066238**

Key species

Mountain Hare.

Access

The small village of Lynemore is on the A939, between Grantown-on-Spey and Bridge of Brown.

Site description

The pasture and heather-clad hills around Lynemore are alive with **Mountain Hares**, and they can be readily viewed from the comfort of the car. There are several lay-bys along this road. Look for the animals on the ridges as they can easily be picked up against the skyline.

23 Findhorn Valley, Highland

Grid ref: NH801289 (Tomatin)

Key species

Red Deer, Sika Deer, Mountain Hare, Feral Goat.

Access

There are various access routes to this lengthy valley but the best is probably from Tomatin. Take the A9 north from Aviemore for about 12 miles (19km) turning left by the Little Chef and then turn right after a couple of miles (3km) down the minor road towards Coignafearn Lodge.

Site description

The spectacular mountainous landscape of the Findhorn Valley offers a dramatic backdrop in which to observe some typical upland mammal species. It should be possible to find good numbers of **Mountain Hares** (80+ seen in March 1993) although be aware that Brown Hares also occur in fields at the start of the valley. **Sika Deer** may be observed early in the morning or late in the evening feeding in the wooded area before the open moorland is reached. A particularly good spot is the woodland along the side road that runs north-west out of Findhorn Valley towards Farr, especially late in the evening. A small herd of **Feral Goats** may also be found in this area. Hundreds of **Red Deer** can normally be seen on the moorland further along the main Findhorn Valley.

24 Monaughty, Moray

Grid ref: NJ160580

Key species

Pine Marten.

Access

From Elgin take the B9010 Dallas road for one mile (1.6km), then the Pluscarden minor road for a further 2.5 miles (4km) and the car park is on the right. There are walks through Forestry Commission woodlands that start off at Torrieston car park (NJ164588).

Site description

The pinewoods of Monaughty contain some magnificent individual Douglas Fir and pine specimens that offer sanctuary for **Pine Martens** as well as more common mammals including **Roe Deer, Red Deer, Weasels, Stoats** and **American Mink**.

25 Culbin, Moray

Grid ref: NH896574
(Nairn east beach car park)

Key species

Red Squirrel, Badger, Roe Deer.

Access

There are three main access points to this forest. The Nairn east beach car park (NH896574) is a large car park managed by the Highland Council from where a signposted path leads along the shore. From Nairn follow signs to the harbour and caravan park and continue through it to the car park. Wellhill car park (NH997614) provides direct access to several way-marked walks, including an all-abilities trail, and is reached by following the Culbin forest signs beyond Kintessack. Alternatively, Cloddymoss car park (NH981599), found by following signs to Cloddymoss from Kintessack, provides linking footpaths to Wellhill.

Site description

This large coastal pine forest, stretching for nine miles (14km) between Nairn and Findhorn Bay, was planted in the 1920s to help stabilise the drifting sand dunes. It is currently managed by the Forestry Commission and supports a large population of **Red Squirrels** which can be seen throughout the forest. There are also several **Badger** setts in the area; these animals take advantage of the soft sandy soil for digging. **Roe Deer** are also present throughout Culbin and are most often seen in the early morning along the trails when they feed on the trackside vegetation.

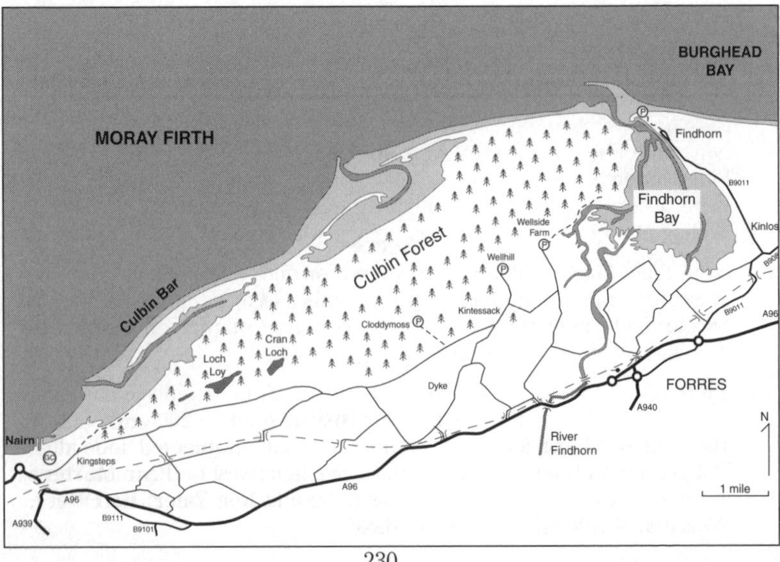

26 Moray Firth and Cromarty Firth, Moray

Grid ref: NH809615

Key species

Bottle-nosed Dolphin, Killer Whale, Minke Whale.

Site description

Bottle-nosed Dolphin: A resident population of up to 130 animals is centred on the Moray and Cromarty Firths. During the winter and spring months many individuals are also found to the south of the Firth, in waters off the coast of Aberdeenshire. During this time, they can frequently be seen off Aberdeen, with some individuals actually entering the harbour at times. The 'Bottlenose season' in this area seems to be from early March through to June. Boat trips during May and June record Bottle-nosed Dolphins on around 70% of trips. Good spots for watching the dolphins in the Moray and Cromarty Firths include Chanonry Point, Burghead, Spey Bay, Fort George, Cromarty and North Kessock.

Killer Whale: In recent years the Moray Firth has attracted Killer Whales during the spring, particularly in March. They are presumably attracted into the Firth by following prey species making inshore movements at this time of year.

The Moray Firth has also started to produce more sightings of **Minke Whales**, particularly in spring, whilst both **Grey Seals** and **Common Seals** have discrete populations in the inner Firth.

27 Glen Urquhart, Highland

Grid ref: NH451458

Key species

Stoat, Pine Marten, Red Squirrel.

The forest is located on both sides of the glen adjacent to the A831 that runs from Drumnadrochit to Corrimony, from the A82 on the western edge of Loch Ness. This scenic glen contains a mix of different habitat types and is home to a diverse selection of woodland mammals. A network of forest roads runs through the glen that are ideal for walking, cycling and horse riding. One of the best places to go in Glen Urquhart is Craigmonie (below).

Craigmonie, Highland

Grid ref: NH503294

Access

There is no car park on site so it is best to park in the village car park, adjacent to the A82 and then follow Pitkerrald road for 200 yards, before taking the left-hand fork onto a farm track, then another left soon after. The woodland is a further 200 yards along here, west of the track. There is a network of walks that link up with the adjacent Balmacaan Wood, owned by the Woodland Trust Scotland.

Site description

This mixed forest of planted conifers and birch is a community woodland, owned by the Forestry Commission but managed in association with the Craigmonie Woodland Association. There are some specimen conifer trees which were planted in the 1800s but which are probably avoided by the local **Pine Martens**. Numerous other species that may be found here include **Stoats, Red Squirrels, Red Foxes, Badgers** and **Roe Deer**.

28 Glen Affric, Highland

Grid ref: NH282282 (Dog Falls car park)

Key species

Sika Deer, Red Deer, Pine Marten.

Access

Glen Affric is five miles (8km) west of Cannich off the A831. There are several car parks in the area and the forest roads provide access to the heart of the Caledonian pinewoods.

Site description

Glen Affric contains the third largest remnant of ancient Caledonian pinewoods in Scotland. It is also one of the least disturbed areas and has been under conservation management since the 1960s. Aside from the pine woodland, the glen also boasts lochs, moorland and mountains. At least 22 mammal species have been identified within Glen Affric.

Sika and Red Deer: Both Sika and Red Deer may be observed grazing on the moorlands and mountainsides within the glen. During colder weather in winter they come down to lower altitudes to forage among the Scot's Pines of this ancient Caledonian forest.

Pine Marten: Pine Martens are the main predators in the area and although they are difficult to spot, they are considered to be 'abundant' and are known to visit Dog Falls car park and picnic area at night to pick up any scraps left by day-trippers. There are several other car parks in the area and the valley road is probably worth driving at night should the visit to Dog Falls car park prove fruitless.

Other mammals: **Wild Cats** are known to occur in Glen Affric but are extremely scarce and difficult to see. Other carnivores present include **Badgers**, **Weasels**, **Stoats** and **Red Foxes**, whilst **Otters** patrol the loch-sides. **Red Squirrels** are also present although they are more difficult to spot at Glen Affric than at some of the other sites listed in the book. **Mountain Hares** occur on the moorland that cloaks the valley sides along much of the glen.

Other information

A good selection of Caledonian pinewood birds can also be found at Glen Affric, including Capercaillie, Crested Tit and Scottish Crossbill. This area is also outstanding for *Odonata* (dragonflies and damselflies), with acid bog species abundant and including White-faced Darter, Northern Emerald, Brilliant Emerald, Highland Darter and Azure Hawker.

29 Plodda, Highland **Grid ref: NH280238**

Key species
Pine Marten, Red Squirrel.

Access
Plodda Forest is approximately 12 miles (19km) west of Inverness near the head of the Beauly Firth. From the village of Beauly go through Tommich and continue along the forest track. Plodda Falls car park is 3.5 miles (5.5km) along here.

Site description
This mixed conifer forest is owned and managed by the Forestry Commission and includes many magnificent, mature conifers interspersed with native broadleaves. **Pine Martens** are found here, but due to their habits, they are not easy to see. Other mammals found in Plodda Forest include **Red Squirrels**, **Roe Deer**, **Red Deer**, **Badgers** and **Otters**.

30 Achnashellach, Highland **Grid ref: NH040493**

Key species
Sika Deer, Red Deer, Roe Deer, Pine Marten.

Access
There are areas of forest on either side of the A890, with a car parking area near the Craig Hostel at the grid reference given above. Several routes and long-distance treks start from the car park.

Site description
Achnashellach is an area of mixed conifer forest managed by the Forestry Commission that cloaks the River Carron valley on both sides. The south side consists mainly of ancient Caledonian pine forest, while the northern slopes have mainly been planted with exotic species. The best time to observe both **Red Deer** and **Sika Deer** at Achnashellach is during winter when they come down to lower altitudes for shelter in bad weather. The deer can either be observed within the pinewoods or grazing on the hillsides. **Roe Deer** also occur but are by habit, trickier to observe. **Pine Martens**, **Badgers** and **Red Foxes** are also present.

Pine Marten (*U. Iff*)

31 Beinn Eighe NNR, Highland

Grid ref: NH000650

Key species

Pine Marten, Otter, Red Deer.

Access

The reserve is two miles (3km) north-west of Kinlochewe off the A832 road. The visitors' centre is open from Easter until October but the trails and car parks can be used all year.

Site description

Beinn Eighe was Britain's first National Nature Reserve and features a wide array of landscapes and habitat types including mountains, ancient Caledonian pinewood fragments and Loch Maree. **Pine Martens** are known to visit the car parks at NH000650 and at NG982670 during the late evening in summer, when they come to search for scraps discarded by day-trippers. **Red Deer** are relatively common and easy to see on the hillsides.

To the south of Kinlochewe is Lochcarron. The loch itself and nearby Loch Kishorn are both good for **Otters** on the falling tide. **Pine Martens** can also be seen most nights at a bed and breakfast in Lochcarron. Telephone 01520 722377 for more details.

32 Kinloch, Highland

Grid ref: NG703161

Key species

Roe Deer, Grey Seal, Common Seal.

Access

Kinloch Forest and car park is on the east side of the A851, four miles (6km) south of Skulamus. The car park is signposted.

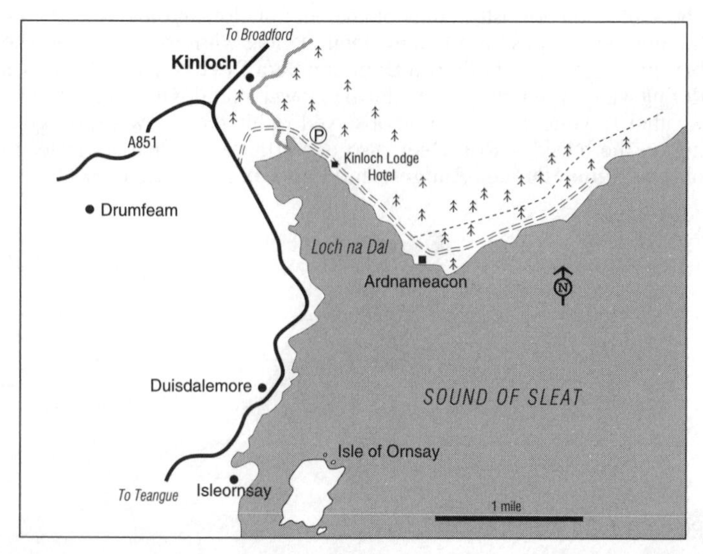

Site description

The woods at Kinloch overlook the Sound of Sleat. They are a mixture of ancient ash-oak on acid soil, parts of which have been designated as a Site of Special Scientific Interest (SSSI), together with patches of coniferous plantation. **Roe Deer** may be spotted anywhere in the area and as the majority of exotic conifers in the wood have been felled to make way for native trees, observing the deer may now be less difficult. Both **Grey** and **Common Seals** may be observed along the coastline of the Sound of Sleat. **Otters** are also occasionally encountered. The area also supports large populations of many small mammals including **Pygmy Shrews**, **Common Shrews**, **Field Voles** and **Wood Mice**.

33 Inner Hebrides and mainland Argyll and Bute

Key species

Harbour Porpoise, Bottle-nosed Dolphin, Common Dolphin, White-beaked Dolphin, Killer Whale, Risso's Dolphin, Humpback Whale, Minke Whale, Common Seal, Feral Goat, Otter.

Site description

The Inner Hebrides are undoubtedly one of Britain's best areas for watching cetaceans. The seas are rich in food and support large populations of numerous species. They are also relatively easy to see thanks to regular boat trips organised by local tour companies (details on pp. 266–267).

Harbour Porpoise: As might be expected, Scotland's most common cetacean species is relatively easy to see in the right locations within the Inner Hebrides. Its most popular haunts are numerous and it can be seen throughout the year in the Sound of Mull, the Sound of Sleet, the Firth of Lorn, off Gairloch, off the Ardnamurchan Peninsula and in the Sound of Jura. Good numbers can be seen throughout most of the year in the waters off the Isle of Mull, but they are sometimes scarcer during the midsummer months. Boat trips during spring usually record porpoises on every trip.

Bottle-nosed Dolphin: The waters surrounding the Isle of Mull are home to a resident population of Bottle-nosed Dolphins, with around 20 individuals in the group. They have proved to be a great attraction for tourists with summer and autumn providing optimum viewing conditions for this particularly exhibitionist school. Dervaig Bay and Tobermory Bay seem to be favoured areas.

The waters around the Isle of Islay are also home to a group of resident Bottle-nosed Dolphins, known as the Ileach dolphins. There are seven in this pod and although they do roam around the Inner Hebrides they are most often seen around the Isle of Islay, which is how they got their name - 'Ileach' means 'of Islay'. They are most often seen during spring and early summer. Caledonian MacBrayne operates the car ferry terminal at Kennacraig (NR825621) on the Kintyre peninsula. Consult the Caledonian MacBrayne website for details of timetables, bookings and fares: www.calmac.org.uk/islay. The waters near Crinan (NR787943) on the Argyll & Bute mainland have recently become another Bottle-nosed Dolphin hot spot. Crinan can be accessed via the B8441 off the A816 north of Lochgilphead.

Common Dolphin: The Inner Hebrides can provide sightings of Common Dolphins throughout the summer until the groups start to disperse around mid-August. As is often the case with this species, schools are often relatively large, numbering up to 120 individuals at times. The waters surrounding the Isle of Mull have traditionally hosted groups of this species, and companies running whale-watching trips in the area usually have a high level of success with this species. Just west of Mull, Common Dolphins often appear in waters around Coll, and they can be viewed from ferries to the isle as well as in the Sound of Mull en route. If you want to stick to the mainland then Point of Ardnamurchan looks west over the waters off the north coast of Mull towards Coll and can be quite productive.

Risso's Dolphin: September and October are the best times to watch Risso's Dolphins in the Inner Hebrides. Their frequent breaching and bold movements can be encountered in waters around Coll and the Small Isles. Often the best chance of successful observation is to join one of the commercial whale-watching vessels that operate in the area. However, for those less keen to venture out onto the open seas, Point of Ardnamurchan on the mainland has produced many sightings in recent years, mainly because of increased observer coverage.

White-beaked Dolphin: The Hebridean Whale and Dolphin Trust (HWDT) has been running a cetacean-observation project from the lighthouse at Point of Ardnamurchan since the spring of 2001. In its first year White-beaked Dolphins were recorded from the end of August into the autumn.

Killer Whale: Killer Whales are observed in the Inner Hebrides fairly regularly. Pods are often encountered around the Small Isles and can sometimes be seen from the commercial whale-watching vessels that operate in the area. The name Ardnamurchan is believed by many to have derived its name from the Killer Whale. Whether this is the case or not, Point of Ardnamurchan is certainly a prime mainland spot from which to try and see the creatures. Sightings generally occur between May and September. Killer Whales often stick close to the coastline, searching for seals and other inshore prey species.

Humpback Whale: In recent times observer effort in Inner Hebridean waters has increased hugely and as a result small numbers of Humpback Whales are now being seen irregularly around the Inner Hebrides and the Isle of Skye. To see this species in this area requires a lot of luck.

Minke Whale: The waters surrounding the Small Isles (Canna, Muck, Rhum and Eigg) and those around the Isle of Mull and Coll provide exciting opportunities to view Minke Whales throughout the summer months. The Hebridean Whale and Dolphin Trust (HWDT) collates data gathered by tour-boat operators and more specialist survey teams (particularly SeaLife surveys) in order to establish population levels and demographic details. In recent times, boat trips off Coll and Mull during May have recorded Minke Whales on over 90% of excursions. Similar –success rates are reported from other areas and on occasion, several animals may be observed from tour-boat excursions.

If you want stay on the mainland, Point of Ardnamurchan, the most westerly point in mainland Britain and possibly the best mainland site for watching cetaceans, is a very good spot from which to watch. The point looks over very productive seas towards the Small Isles, where Minke Whales are regularly recorded. There is a small watchpoint especially built for cetacean recording particularly when the weather is less than ideal.

Common Seal: The south-east coast of the Isle of Islay holds a nationally important population of Common Seals. Islay is used extensively for pupping, moulting and as a hauling-out site. The animals can be easily viewed anywhere on the island. See 'Bottle-nosed Dolphin' for details of how to get to Islay.

Otter: Rhum provides the full range of marine and freshwater requirements necessary for Otters, including breeding, feeding and resting sites. There are numerous freshwater lochs and lochans at a range of altitudes from near sea level to over 400m, a range of small rivers and streams and an extensive coastline. Good areas to try include the beaches at Kinloch (NM405995) and where the Kilmory River meets the sea at the north end of the island (NM361040). Otters are also relatively easy to observe on the Isle of Mull: indeed, it has one of the densest populations of Otters within the UK. They are widespread in all of the sea lochs and can be observed from many of the coast roads around the isle.

Rhum is also home to a population of **Feral Goats**. The population has fluctuated between 98 and 185 individuals. A pure population of **Red Deer** also occurs on Rhum and has been the subject of numerous scientific papers regarding animal behaviour and population dynamics. **Common Seals** are also found here. There are year-round daily sailings to Rhum and the Small Isles with Caledonian Macbrayne: www.calmac.co.uk.

Further details on mammal watching in the Ardnamurchan Peninsula can be found under that site entry.

34 Ardnamurchan Peninsula, Argyll and Bute

Key species

Wild Cat, Otter, American Mink, Red Deer, Roe Deer, Pine Marten, Grey Seal, Common Seal.

Access

The Ardnamurchan Peninsula is one of the best areas for watching mammals in Britain. With a bit of luck Wildcats, Pine Martens, Otters and a

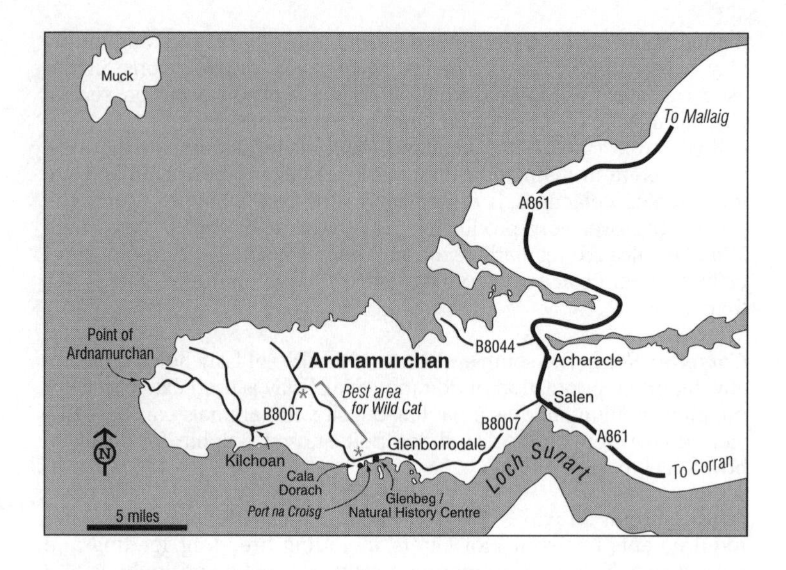

diverse selection of other species may be seen. The peninsula is best reached via the Corran-Ardgour ferry, off the A82, 13 miles (21 kms) south of Fort William.

Site description

Grey and Common Seals: Both of these are relatively easy to observe on the peninsula. Good spots include Loch Linne, south-west of Ardgour, where both may be observed loafing on rocks along the adjacent stretch of road; Loch Sunart between Ardnastang and Glenborrodale; Port na Croisg, a bay just west of the Natural History Centre at Glenmore; and the Point of Ardnamurchan (NM415675), where they are particularly common and easy to observe.

Wild Cat: The 5–6 mile (8–9.5km) stretch of road from Cala Darach to the junction with the Kilmory road (at NM526665) is reputedly the best area for Wild Cats. The best stretch is the two miles (3km) or so beyond Cladh Chiarain, especially the scree slopes (at NM560623) and the rush fields (at NM553630) where they hunt for voles. The best chance of seeing one is to go spotlighting an hour or so after dark but they are by no means easy to find. Most people have needed to spend at least two nights spotlighting before being successful. A powerful spotlight is essential as the cats may be well off the road. Wild Cats have also been seen along the road from Kilchoan to Sanna, at the western end of the peninsula.

Otter: Otters are widespread and reasonably common across the peninsula and could be found in any area of likely looking habitat. Good areas to try include the eastern end of Loch Sunart, just east of Strontian, and Port na Croisg, a bay just west of the Natural History Centre at Glenmore. Although dawn and dusk are the prime times for Otters, they can be seen throughout the day at this latter site, especially on the falling tide as the rocks and pools become exposed. Otters also frequently occur in the bay opposite Cala Darach and off the Point of Ardnarmuchan, particularly in Briaghlann Bay to the south of the point.

Cetaceans: In recent times an array of cetaceans have been noted utlising the rich seas of the Ardnamurchan Peninsula for feeding. **Minke Whales, Common Dolphins** and **Harbour Porpoises** are all possible off the Point of Ardnamurchan, particularly during summer, while **Harbour Porpoises** are frequent in Loch Sunart as far east as Salen. See 'Inner Hebrides' for further and more comprehensive details on cetacean watching in the area.

Pine Marten: Pine Martens are common on the peninsula, particularly along the edge of Loch Sunart where they visit bird tables at houses along the road. Spotlighting along this section of the B8007 should guarantee sightings. They are also frequently seen if you spotlight along the road between Cala Darach and Cladh Chiarain (NM562617).

Bats: Although the diversity is low, there are still a couple of species of bat worth keeping an eye out for if visiting Ardnarmurchan. **Brown Long-eared Bats** occur at Glenmore Natural History Centre, while **Common Pipistrelles** breed in the roof of Cala Darach, a former bed and breakfast just west of Glenmore (NM578618).

Other mammal species

Water Shrews occur on the shore of Loch Sunart east of Salen.
American Mink occur commonly along Loch Sunart.
Both **Roe Deer** and **Red Deer** may be seen just south of the Natural History Centre on Eilean Mor, while the moorland further along the road past Loch Mudle and just before the junction with the Kilmory road is excellent for **Red Deer**: well over 100 can be seen here, principally from dusk to dawn. Fewer are seen during the day.

Further details on mammal watching in the Inner Hebrides and Ardnamurchan Peninsula are given under the 'Inner Hebrides' site entry.

35 Isle of Skye, Highland

Grid ref: NG125465 (Neist Point)
NG217598 (Ardmore Point)

Key species

Killer Whale, Atlantic White-sided Dolphin, Minke Whale, Common Seal, Otter.

Access

The Isle of Skye is connected to the mainland by a short road bridge, the Skye Bridge, just east of Kyle of Lochalsh.

Site description

Atlantic White-sided Dolphin: In recent years records of Atlantic White-sided Dolphins have increased in the seas around the Isle of Skye, with Loch Ainort being a favoured haunt. Neist Point and Ardmore Point are prominent headlands that offer good sea-watching opportunities. Late summer is a good time to watch for this species.

Killer Whale: Killer Whales are often encountered by boats in the Minch, the area of sea between the mainland and the Isle of Lewis. Western and northern points of the Isle of Skye overlook this area and many records of this species have come from these points in recent years. As well as Ardmore Point and Neist Point, the headland of Borneskitaig offers good opportunities for seeing this species.

Minke Whale: Any prominent headland on the Isle of Skye provides the opportunity for potential viewing of Minke Whales in the summer months.

Common Seal – Ascrib, Isay and Dunvegan: This site, consisting of islets, offshore islands and mainland shores holds around 2% of the UK Common Seal population. They can be viewed around the coast of Skye. Suitable promontories include Ardmore Point and Geary.

Otters: Otters are common around the coast of the Isle of Skye. A particularly good spot is Kylerhea (Kile-ray) Otter Haven, seven miles (11km) off the A89. Take signposts for Kylerhea and the Otter Haven is signposted from the village. From the car park follow signs along the forest track to the observation hide. This overlooks beautiful scenery, including the shoreline, ponds and rivers, which are the habitat of several resident Otters. The Otters are active during the day year-round, and **Red Deer**, Golden Eagles, seals and seabirds are also present. The Haven is open from 09:00 hrs until one hour before dusk all year. For more information contact the Fort Augustus Forest Enterprise Office on 01320 366322.

36 Outer Hebrides

Key species

Risso's Dolphin, Atlantic White-sided Dolphin, Grey Seal, Black Rat, Otter.

Access

The Outer Hebrides are best accessed by ferry. Caledonian MacBrayne runs car and passenger ferries to all the islands all year round: Ullapool to Stornoway (Lewis); Uig (Skye) to Tarbert (Harris) and Lochmaddy (North Uist); and Oban to Barra. Ferries also run from Castlebay (Barra) to Lochboisdale (South Uist); and Berneray (North Uist) to Leverburgh (Harris). A car ferry now runs daily between Barra and Eriskay. You can check timetables and book online at www.calmac.co.uk. Inter-island ferries run from Berneray (North Uist) to Leverburgh (Harris); Tarbert (Harris) to Lochmaddy (North Uist); and Castlebay (Barra) to Lochboisdale (South Uist). Highland Airways operate daily flights (Monday–Friday) from Stornoway (Lewis) to Benbecula, while British Airways flies between Stornoway and Barra daily (depending on the tide as the plane lands on the beach).

Site description

Risso's Dolphin: The seas surrounding the Outer Hebrides are a good area to look for Risso's Dolphins. Favoured haunts are around the Isle of Lewis, and this species is often encountered from the ferry that runs through the Minch from Ullapool to Stornoway, Lewis. They are semi-resident off Tiumpan Head (NB574380) on Lewis. The optimum time is late summer, usually from around mid-July until the end of October.

Atlantic White-sided Dolphin: Atlantic White-sided Dolphins are sporadically recorded from the Outer Hebrides. Prominent headlands on the northern, western or eastern coastlines, particularly Tiumpan Head on Lewis, give a chance, while the ferry to Stornoway from Ullapool across the Minch crosses a traditional area for this species.

Bottle-nosed Dolphin: One of the few reliable spots on the Outer Hebrides for Bottle-nosed Dolphins is the Sound of Eriskay off the southern coast of South Uist. They can be seen there throughout the year.

Other cetaceans

White-beaked Dolphins used to be frequently encountered, often in large numbers off Tiumpan Head on Lewis, although recent research suggests a dramatic decline in the occurrence of this species in the Outer Hebrides possibly as a result of increasing sea temperatures. **Minke Whales** may also be seen from this watch point, as deep water surrounds the head on all sides, providing ideal feeding conditions for this species. Records of **Killer Whales** are annual from the Outer Hebrides, particularly from the Isle of Lewis, where Tiumpan Head and the Butt of Lewis produce the majority of sightings.

Grey Seal: At the beginning of 2001, the Outer Hebrides had a total Grey Seal population of 38,000 individuals, approximately 40% of the world population. They breed on offshore islands such as the Monarch Islands and Shillay. The Monarch ('Monk') Islands provide a wide area of largely undisturbed habitat for breeding Grey Seals. These islands hold the largest breeding colony in the UK, contributing over 20% of the UK's annual pup production. The islands, managed by Scottish Natural Heritage, are difficult to access, and the site is not recommended to the casual observer. However, Grey Seals do wander widely while feeding and outside the breeding season they can be encountered off any point of the Outer Hebridean coastline, while fishing ports such as Stornoway (Lewis) usually have a few individuals loafing around looking for scraps and hand-outs. **Common Seals** may also be encountered almost anywhere around the islands, with the south coast of North Uist being a favoured feeding area.

Black Rat: The small archipelago of the Shiant Islands (pronounced 'Shant') in the Minch, between the Isle of Harris and the Isle of Skye, is home to a small population of Black Rats. They occupy cliff habitats and rocky outcrops on Garbh Eilean, the main island of the archipelago. A mark-recapture study in May 1996 revealed there to be between 230 and 400 individuals on the islands. It is felt that the comparatively small rat population has little effect on the large seabird colonies on the islands. Boat trips are available from Tarbert during the summer months.

Other species

Red Deer are fairly common on the Outer Hebrides where they are actually an introduced species. There are just over 4,000 on Lewis and Harris, around 850 on North Uist and 350 on South Uist. They are regularly seen from the road around Langass in North Uist, particularly at night.

Otters are relatively common on the Outer Hebrides. Particularly good areas include:

- Loch Langass on North Uist, where they can regularly be seen about half a mile from the Langass Hotel along the footpath starting behind the hotel. The hotel organises regular 'Otter Walks'.
- The bay viewable from the Lochboisdale Hotel (South Uist).
- Around Lochmaddy and Loch an Duin (North Uist), where many holts can be found along the sheltered coastline.

American Mink have colonised Lewis and Harris and are threatening many of the ground-nesting seabird colonies on the islands. The 'Hebridean Mink Project' is currently looking at ways to try to stop this species spreading further south to the island of North Uist.

Wood Mice occur in three different forms on the Outer Hebrides: one on St Kilda, a Hebridean variety and one that exists at the southern tip of the Outer Hebridean chain.

Rabbits are common throughout the islands as are small populations of introduced **Mountain Hares**. The recently introduced **Hedgehog** populations have become a threat to many ground-nesting birds and plans are afoot to try either to cull the population or translocate the animals back to the mainland.

Mountain Hare (*James Gilroy*)

37 Dornoch Firth, Highland

Grid ref: NH805873 (Dornoch Point)

Key species

Common Seal.

Access

A particularly good place to watch the Common Seals at this site is from Dornoch Point, accessed south from Dornoch village along a permissive footpath past the disused airstrip and golf club. The point looks west over Dornoch Sands and east over Gizzen Briggs.

Site description

A significant proportion of the Inner Moray Firth **Common Seal** population resides within Dornoch Firth, the most northerly large estuary in Britain. The seals, which utilise sandbars and shores at the mouth of the estuary for haul-outs and breeding sites, are the most northerly population to use sand banks. Their numbers represent around 2% of the UK population.

The River Evelix, which enters the Firth approximately three miles (nearly 5km) west of Dornoch, supports a sizeable population of **Otters**.

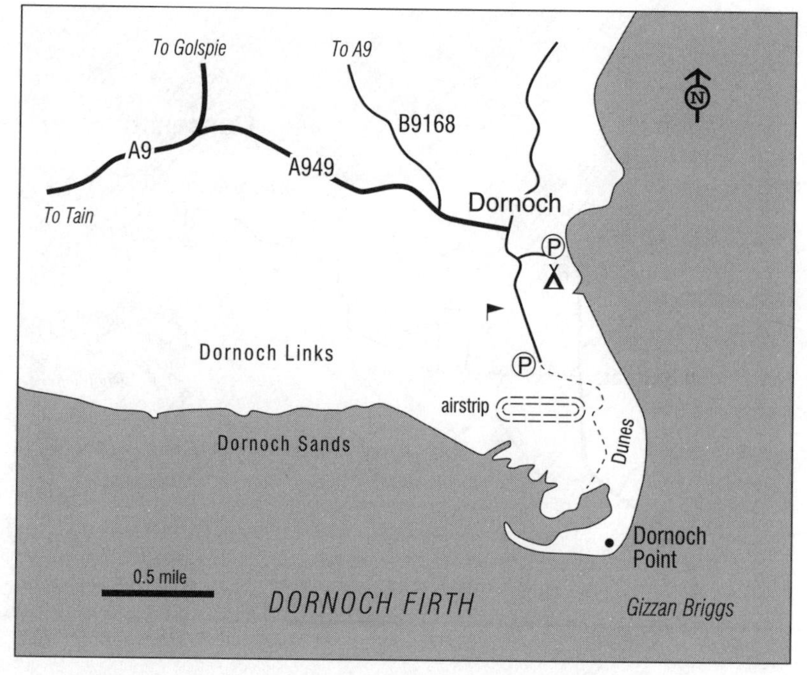

38 Duncansby Head, Highland

Grid ref: ND403733

Key species

Wild Cat.

Access

Duncansby Head to the east of John O'Groats is the most north-eastern point of mainland Britain and is accessible from Wick or Thurso.

Site description

Park at the lighthouse and walk along the clifftop heading west. You will come to a small gully running inland from the clifftop. In this gully, which is only around three metres deep there is a square hole in the rock face on the western side. An adult **Wild Cat** was photographed in this hole in 1995 and fresh droppings were found here the following year.

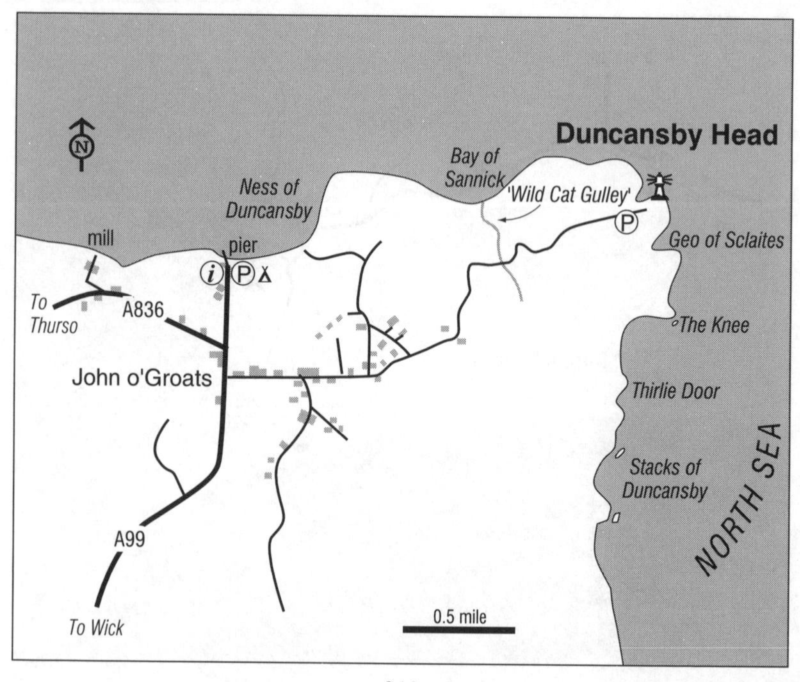

39 Orkney

Key species

Killer Whale, Grey Seal, Common Seal, Common Vole.

Access

Orkney can be accessed by ferry or plane. There are direct flights to Kirkwall airport on Mainland Orkney daily (except Sundays) from Aberdeen, Wick, Edinburgh, Glasgow, Inverness and Shetland. All these flights are operated by Loganair/British Regional Airlines and can be booked through British Airways.

There are several ferry routes to Orkney from the mainland. Northlink ferries operate from Scrabster and Aberdeen to Stromness on a regular basis, daily on the Scrabster route. See their website for the current timetable: www.northlinkferries.co.uk. The Aberdeen ferry often offers good opportunities to see cetaceans.

John o' Groats Ferries operate a passenger-only (and bicycles) ferry service from John o' Groats to Burwick (two to four times per day; 45 minutes) on South Ronaldsay, from May to September.

Once there, getting around is fairly simple with numerous inter-island ferries and flights operating to and from most of the islands.

Site description

Killer Whale: The summer months produce frequent sightings of Killer Whales around the Orkney Islands, possibly involving transient pods heading south from the Shetland Islands. Any spot can produce sightings, but if you want to increase your chances then heading to North Ronaldsay, the archipelago's most northern isle, would be a good idea. In the past few summers pods of up to ten have been seen from North Ronaldsay. Access to the island is possible by boat or plane.

Grey Seal: The two uninhabited islands of Faray and Holm of Faray in the northern part of Orkney support a well-established Grey Seal breeding colony. The seals tend to be found where there is easy access from the shore, and freshwater pools on islands appear to be particularly important. These islands support the second largest breeding colony of Grey Seals in the UK. At the start of 2001, Orkney held 54,000 Grey Seals. Visits to Faray and Holm of Faray are possible by local arrangement.

Common Seal: Situated in the north-east of the archipelago, Sanday supports the largest group of Common Seals in Scotland. The breeding groups, found on intertidal haul-out sites that are unevenly distributed around the coast of Sanday, represent over 4% of the UK population. Inshore kelp-beds surrounding Sanday are important foraging areas for the seals. Ferries to Sanday depart from Kirkwall on Mainland Orkney and take one hour and 25 minutes.

Common Vole: Neolithic settlers probably introduced the Common Vole to Orkney. Since then, it has done rather well: the population on the whole of Orkney is estimated to number around 1,000,000 individuals. As well as occurring on mainland Orkney, the Common Vole can be found on Westray, Sanday, Stronsay, South Ronaldsay and Rousay. It inhabits a wide range of habitat types including coniferous plantations, deciduous plantations, marshland, heather moorland, hay meadows, ditches and gardens. It tends to avoid short pasture and arable land.

40 Shetland

Key species

Harbour Porpoise, Risso's Dolphin, White-beaked Dolphin, Atlantic White-sided Dolphin, Killer Whale, Fin Whale, Humpback Whale, Minke Whale, Common Seal, Grey Seal, Otter, Mountain Hare.

Access

There are daily ferries to and from Aberdeen (www.northlinkferries.co.uk). The crossing normally takes around 12 hours and it is possible to take your car on board. Cetaceans are often seen from the ferry. Alternatively Sumburgh is Scotland's fourth busiest airport. British Airways offers flights from Aberdeen (one hour), Edinburgh (two hours), Inverness, Wick and Glasgow.

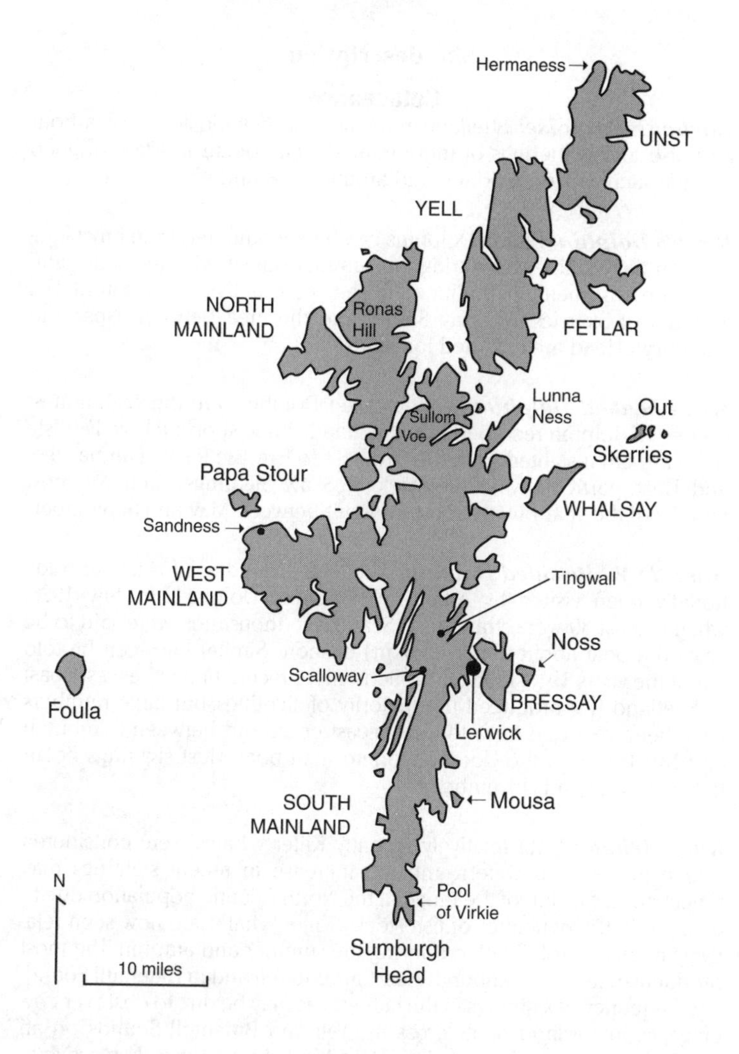

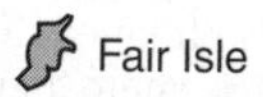

Getting around Shetland is also relatively easy as there are frequent inter-island car ferries between Mainland Shetland and the outlying islands. Scheduled bus services operate throughout the mainland and some of the larger islands. For further information on getting to and around Shetland go to www.shetlandtourism.com.

Site description

Cetaceans

Harbour Porpoise: Shetland remains a stronghold for the Harbour Porpoise, and gatherings of more than 100 can occur in areas of good feeding such as Mousa Sound and around Noss and Whalsay.

Risso's Dolphin: Risso's Dolphins can be encountered from any headland on the archipelago during the summer, but sites on the south and east sides of Shetland traditionally see more activity. Baltasound, Yell Sound, Out Skerries, Whalsay Sound and the area between Noss and Sumburgh Head are favoured localities.

White-beaked Dolphin: White-beaked Dolphins are the commonest species of dolphin recorded from Shetland. Prime spots include Fair Isle (they are often sighted from the Grutness to Fair Isle ferry), Lamba Ness and Unst, particularly at the Wick of Skaw. Sightings occur in most months of the year but are most frequent between May and September.

Atlantic White-sided Dolphin: The area around Scalloway has traditionally been visited by Atlantic White-sided Dolphins. In July 1926, when at least 30 were stranded at Scalloway, 'thousands' were said to be around a boat about a mile (1.6km) offshore. Similar tales can be told about the years 1919, 1936 and 1990. In more recent times, the west coast of Shetland has produced the majority of sightings, but large numbers have been reported from the north-east coast and between Sumburgh and Fair Isle from the Good Shepherd mail boat. Most sightings occur between June and November.

Killer Whale: Until relatively recently Killer Whales were considered scarce in Shetland waters, and the increase in recent sightings may reflect an expansion of the range of the North Atlantic population due to changes in the dynamics of fish stocks. Killer Whales are now seen relatively frequently off Shetland during the summer and autumn. The most productive sites are Sumburgh Head, around Yell and in Bluemull Sound. The frequency of sightings in the latter areas may be due to observer coverage, as inter-island boats cross the Yell and Bluemull Sounds on an hourly basis. The pods are transient and it is thought that they arrive in Shetland from the north, passing through the Sounds in the North Isles, following the east coast, passing Noss, Mousa Sound and Sumburgh Head. The peak time for sightings is usually in early June and July. Pods normally consist of up to 12 animals with groups of 4–6 being most common, although sightings of 1–3 are also frequent. They often spend days or weeks at a time around seal colonies where they take both adults and pups, but they also prey on shoaling fish and small cetaceans.

Fin Whale: Over 4,300 Fin Whales were taken in Shetland waters between 1903–1914 and 1920–1929, many being caught off the edge of the continental shelf in the Faeroe Channel north of the islands. Unfortunately, at present, this species is rarely recorded in Shetland waters. However, small numbers continue to be seen in the Faeroe Channel during pelagic cruises, with most records coming between April and October, particularly from May to August.

Humpback Whale: Between 1903 and 1914, a total of 49 Humpback Whales were taken by whalers in Shetland waters. This number reduced to just two animals being taken between 1920 and 1929. Since then there have been a few records and Humpbacks were noted off Sumburgh Head in June 1992, May 1993, September 1993, and throughout the summers of 1994, 1995, 1996, 1997, 1998 and 1999. Thereafter Humpbacks have been recorded less frequently, but individuals are still observed in Shetland waters during the summer. However, given that they have stopped summering in favoured areas, it is less easy to predict where they will occur.

Minke Whale: Minke Whales are frequently observed in Shetland coastal waters. They are most common on the east side of Shetland where individuals can be seen close to many headlands and small islands. Most sightings occur between April and October, and particularly July to September.

Seals

Common Seal: Situated just half a mile off the east coast of Mainland Shetland, the exposed and uninhabited rocky island of Mousa supports one of the largest groups of **Common Seals** in Shetland. The large, rocky tidal pools on the island are of particular importance, as they are frequently used by the seals for pupping, breeding and moulting, and they provide shelter from the exposed conditions of the open coast. The seals utilise three main areas on Mousa for hauling-out: a steep, flat rock sloping into the sea on the north-west side, and tidal pools on the west and east coasts. The number of seals hauling out generally increases throughout the summer. The site supports just over 1% of the total UK population.

Grey Seals (*E. Hazebroek*)

Grey Seals are also present on Mousa and these can generally be found lying on the water's edge or on offshore rocks. Mousa can be accessed by ferry from Leebotten on Mainland Shetland between mid-May and mid-September.

Grey Seal: The Shetland population of around 3,500 animals represents just over 3% of the British population and just over 1% of the world population. Although Grey Seals may be seen throughout the archipelago, there are few major gatherings, other than seasonal congregations at traditional breeding and moult sites, most of which are well offshore and difficult to access. However, the northern section of Lerwick Harbour is now populated by large numbers of Grey Seals, which gather for free food from the fishermen and Fresh Catch factory workers. Many of these animals are now hand-fed by fishermen and tourists alike. Numbers around the Fresh Catch factory have increased to around 70 animals, with smaller numbers around the salmon-processing factory to the north. The sex ratio here is 9:1, males to females.

Otter: The Otter population on Shetland is considered by many experts to be both morphologically and genetically distinct. It may also be the densest population in Europe although it fluctuates greatly, possibly in relation to changes in the availability of suitable prey. The Yell Sound area has the highest density of Otters and is believed to support more than 2% of the entire British population. Good spots on Yell include Mid Yell Voe, south Yell and Whale Firth. Other good sites on Shetland for viewing Otters include Kirkabister (Bressay), Lunna Ness and Fora Ness (both north-east Mainland).

Mountain Hare: Introduced populations occur throughout Mainland Shetland.

IRELAND

Bottle-nosed Dolphin: In Ireland Bottle-nosed Dolphins are most commonly sighted off the southern and western coasts where small resident or semi-resident populations occur. They may be seen throughout the year although there is usually a drop-off in records between December and April.

A resident population of Bottle-nosed Dolphins resides in the waters of the Shannon Estuary, County Clare. A study in 1994 showed there to be some 56–68 animals present. Favoured areas include Kilcredaun Head, Beal Bar and off Leck Point. May to August is the peak time for viewing these animals in the Shannon, with dolphins being present on over 75% of days each month. It is possible to take boat trips out to view them.

Co. Kerry. The Dingle Peninsula produces many sightings of Bottle-nosed Dolphins with the best mainland spots being Clogher Head, Ventry, Smerwick Harbour, Dingle Bay and Slea Head. There is also a semi-tame resident dolphin that spends its time in Dingle Harbour and who goes by the name of Fungi. He has been in the harbour for over 15 years now and boat trips are available to go and see him.

Co. Cork. Good numbers of this species are seen throughout the year with the best location being Cork Harbour, although sightings are widespread.

Co. Clare. Doolin, Liscannor Bay, Finnis and Loop Head produce the majority of sightings.

Good numbers of Bottle-nosed Dolphins are also recorded off Mayo, Galway and Donegal, with smaller numbers seen irregularly off Sligo, Derry, Antrim and Waterford.

Common Dolphin: As is the case with most cetacean species in Ireland, Common Dolphins are most commonly seen off prominent headlands in Co. Cork. Analysis of sightings of this species from 2001–2005 has shown that Galley Head, Old Head of Kinsale, Cape Clear Island, Dursey Island and Cork Harbour produce the most sightings. They can also often be encountered from points in Co. Kerry (try Slea Head, Bray Head or Sybil Head) and Co. Clare (Spanish Point or Loop Head). Outside of these areas sightings are less frequent.

Risso's Dolphin: During the last 15 years sightings of Risso's Dolphin off Ireland have come from a wide selection of localities spread mainly around the south and west coasts. Co. Cork and Co. Wicklow seem to be the most favoured areas although records regularly come from Wexford, Mayo, Wicklow, Donegal and Kerry. Away from these coasts sightings are less frequent although they have been reported from Dublin and Antrim. The most reliable areas in Co. Wicklow are Wicklow Head, Greystones, Bray Head and Kilcoole. Top spots in Co. Cork include Dursey Island, Galley Head, Cape Clear Island and the Old Head of Kinsale. Dursey Island would appear to be Ireland's prime site with regular sightings of up to 16 individuals.

White-beaked Dolphin: The White-beaked Dolphin is very infrequently recorded around the Irish coastline, with the Irish Whale and Dolphin Group (IWDG) listing only 22 sightings during 1996–2005. No one area seems particularly favoured by this species, although they seem to avoid the Irish Sea as there are no reported sightings from the east coast. The north-west region and Co. Kerry claim the majority of observations with most animals being recorded between May and September.

Atlantic White-sided Dolphin: The Irish Whale and Dolphin Group (IWDG) records only 24 incidences of Atlantic White-sided Dolphins since 1993, with many of these relating to the same pod which loitered for a few days off the north-west coast in July 1996. All sightings have been from the south and west coasts. They have been recorded between March and November with most occurring during high summer.

Killer Whale: Analysis of Killer Whale records off Ireland between 1992 and May 2004 indicate that this species is relatively rare in Irish seas with just 80 reported sightings. This total does not include the famous threesome that spent a week in Cork Harbour during June 2001. They are generally seen off the south and west coasts with sightings ranging from Co. Antrim to Co. Wexford and from all points in between. No location can be said to have regular sightings although prominent headlands and islands in Co. Cork and Co. Kerry have most records. Summer sightings predominate, although observations in the midwinter months of December and January are not unprecedented.

Fin Whale: The Irish Whale and Dolphin Group has established a cetacean watching scheme off Co. Cork based at the Old Head of Kinsale. Fin Whales off Co. Cork start to be recorded in May and numbers slowly build up through the summer. Watches from Co. Cork have revealed a clear late autumn/early winter peak in Fin Whale activity, which seems to drop off by the end of February. This trend has been observed each year since 1999. In 2001 large baleen whales appeared off Co. Cork in June, with a peak of 16–20 Fin Whales seen off Old Head of Kinsale in late July. During November and December Fin Whales were seen on 100% of watches, feeding close inshore from several headlands between Glandore and Old Head of Kinsale.

During 2004, small groups of Fin Whales were seen off Galley Head during the summer and continued to be present off Cork, in groups of up to 30 animals, during that winter, before a group stayed off Ardmore Head, Co. Waterford, for the first two months of 2005. March and April in 2005 again saw a lack of Fin Whale activity from the south coast with the first of the summer being spotted at the beginning of May. Numbers built up very slowly with animals being seen along the south coast throughout the summer and autumn, with whales still being noted to the end of February 2006. At least 83 individuals were recorded in December 2005 and January 2006. Sites in Co. Waterford, particularly Ardmore Head and Helvic Head, have become as reliable as sites in Co. Cork. Small numbers of Fin Whales have started to appear off Slea Head, Co. Kerry during autumn. They are best looked for during calm weather in September.

Humpback Whale: In recent years, Humpback Whales have appeared off the coast of southern Ireland. Pods of up to five individuals have appeared off Galley Head, the Old Head of Kinsale and Cape Clear Island. Humpbacks are not seen off the Irish coast before the end of July, and sightings peter out by the New Year. In the winter of 2001 Humpback

Whale sightings became more regular with up to five individuals being seen along a 20-mile (32km) stretch of coast between Glandore and Kinsale, and a mother and calf were also seen together on three occasions in the week before Christmas in 2001. In 2003 they were recorded from July up until the end of December but with just five sightings in total, while in 2004 up to three individuals again appeared off Cork from November before moving to Ardmore Head, Co. Waterford, early in 2005 where they associated with Fin Whales. In recent years small numbers of Humpback Whales have also been recorded off Co. Clare (e.g. Loop Head), Co. Waterford (e.g. Ardmore Head) and Co. Wexford (Hook Head).

Minke Whale: Minke Whales are relatively common off some parts of Ireland and have been recorded in small numbers off every coast. Co. Cork is undoubtedly the best area in Ireland for Minke Whale sightings. Indeed, they appear to be common and are frequently observed off headlands such as Galley Head and Old Head of Kinsale, as well as off the islands of Dursey and Cape Clear. Minke Whales are most common during the summer months and have been more or less absent between November and March for the last two years. Outside Co. Cork, Loop Head in Co. Clare and Slea Head, Co. Kerry, are productive sites.

Harbour Porpoise: This species is extremely common off Ireland, with large numbers off south-western Ireland, particularly Co. Cork and southern Co. Kerry. Co. Dublin also produces a lot of sightings, good spots being Dun Laoghaire, Dingle Bay and Killiney Bay. Sightings occur throughout the year, although there is generally a drop-off in numbers from January to March.

Lesser Horseshoe Bat: This is the only member of the horseshoe bat family to be found in Ireland, where it frequents the west of the country. Summer colonies can consist of many hundreds of individuals and are usually found in large old houses and farm outhouses, such as barns and stables. The population is currently estimated to be around 12,000, and large nursery roosts are still being discovered. This species may readily be seen at the following sites: Glengarriff Wood, Co. Cork, where **Common Pipistrelles**, **Soprano Pipistrelles**, **Brown Long-eared Bats**, **Leisler's Bats** and **Natterer's Bats** also occur; Killarney National Park, Co. Kerry, and Coole Park, Gort, Co. Galway, where **Common Pipistrelles**, **Soprano Pipistrelles**, **Brown Long-eared Bats**, **Leisler's Bats** and **Natterer's Bats** may also be encountered.

Nathusius's Pipistrelle: A breeding roost of over 160 individuals was found in Antrim (Co. Antrim) in 1997. At the start of 2000 there were known to be over 300 individuals residing in the various buildings of the Clotworthy Arts Centre, Randalstown Road, Antrim. Other species present here include **Soprano Pipistrelles**, **Common Pipistrelles**, **Daubenton's Bats** and **Leisler's Bats.** A colony of Nathusius's Pipistrelles has also been found at Ardress House (a National Trust property), near Portadown in Co. Armagh (grid ref: H918560). Other species present at this site include **Leisler's Bats, Daubenton's Bats** and occasional **Whiskered Bats**. Other individuals have also been recorded in Co. Laois, Co. Derry, Co. Down and Co. Wicklow.

Leisler's Bat: Leisler's Bat is relatively common in Ireland, in fact it is Ireland's fourth-commonest bat species (after both pipistrelles and the Brown Long-eared Bat), and identification is aided because Noctules are not present in the country. This species can be seen at the following sites: The Lough in Cork City (Co. Cork) where **Common Pipistrelles**, **Soprano Pipistrelles**, **Brown Long-eared Bats** and **Daubenton's Bats** may also be encountered; Sandymount Strand, Co. Dublin; Glengariff Wood, Co. Cork; Coole Park, Gort, Co. Galway; Killarney National Park, Co. Kerry; and the Wicklow Mountains National Park, Co. Wicklow, where **Common Pipistrelles**, **Soprano Pipistrelles**, **Brown Long-eared Bats**, **Natterer's Bats**, **Whiskered Bats** and **Brandt's Bats** also occur.

Red Squirrel: Red Squirrels are fairly widespread in Ireland, although numbers are limited along the western and northern coasts. It is felt that this species' origins in Ireland stem from introductions. Indeed, it became extinct in the country early in the 18th century. Current populations are the result of ten introductions of UK animals between 1815 and 1856.

Mountain (Irish) Hare: Mountain Hares of the Irish form, *Lepus timidus hibernicus*, can be found throughout the country at all altitudes. They used to be common but have suffered a decline in the last century possibly due to the conversion of their favoured habitat types into more intensive agricultural land. Certainly, they appear to require hedgerows and species-rich pasture for shelter and resting areas. There are estimated to be 8,250 to 21,000 animals in Northern Ireland alone.

Otter: The Otter is a relatively common species in Ireland and a survey in 1980–1981 found Otter signs at 92% of surveyed sites.

Pine Marten: Pine Martens can be found in many Irish counties and in certain areas they appear to be quite common. Their distribution stretches all the way down the west coast, and they appear to be particularly common in central counties. They are more sparsely distributed down the eastern seaboard.

Black Rat: The Black Rat is extremely rare in Ireland but there have been reports from Co. Armagh, Co. Down and Lambay Island, off the coast of Co. Dublin.

The Badger Trust

Where to Watch Badgers

Avon

Folley Farm Nature Reserve, Stowey, Nr Bristol.
Tel: Avon Wildlife Trust: 0117 917 7270

AWT reserve with well-marked trails, which bypass Badger sett. Located off A368.

Cheshire

Brian Rhodes, Wirral and Cheshire Badger Group.
Tel: 01925 656188

The Badger group takes people Badger watching at a sett in the Frodsham area, in woodland owned by the group.

Cornwall

David and Sarah Chapman
Bosence Meadow
Tel: 01736 850287
www.ruralimages.freeserve.co.uk

The Badger Watch at Bosence Meadow offers more than just a Badger Watch, giving a complete tour of the smallholding. You will also have the chance to see lizards, Adders, Slow Worms, Cetti's Warblers, wildflowers, butterflies and pond life.

Glenville Bed & Breakfast
Pat & Paul Calcraft, Grenville, 1 Quilver Close, Gorran Haven, St Austell PL26 6JT
Tel: 01726 843243
www.cornish-riviera.co.uk

Spectacular sea views and Badgers visiting every evening. Situated in a conservation area, on a hillside overlooking the bay towards Turbot Point. village of Gorran Haven and the Accommodation in a double room or twin room. No children, smoking or pets.

Trevigue Wildlife Conservation
Crackington Haven
Tel: 01840 230418
www.wild-trevigue.co.uk

Badger watch/walk. Self-catering and B&B available at National Trust Farm.

Devon

Paignton Zoo Environmental Park
Tel: 01803 697500

Badger watch by artificial moonlight. Two-hour viewing of Badgers which live naturally in the park. There are now three clans of Badgers. Phone for times and dates. Adults £6.50, PAWS members (inc. children) £4.

Devon

Devon Badger Watch
Tel: 01398 351506
www.devonbadgerwatch.co.uk/

Please make all bookings by telephone. B&B available 5 miles (8km) north of Tiverton. There is a hide and live CCTV pictures during the summer.

Dorset

Badger and Wildlife Watch
Old Henley Farm, Buckland Newton, Dorchester, Dorset DT2 7BL
Tel/Fax: 01300 345293
www.badgerwatchdorset.co.uk

'From our hide enjoy the unique sight of Badgers emerging from their setts and foraging for food. The area is floodlit for perfect vision, and visits from owls, foxes, rabbits and bats add to the spectacle, which is available year-round. Bring a torch and some refreshments and stay until dawn if you want to.'

East Sussex

Bill and Pamela Waghorn, Fantail Cottage, Rosemary Lane, Pett, Hastings TN35 4EB.

Bed and breakfast, with a floodlit patio, regularly visited by Badgers.

Gloucestershire

Tony Dean, Gloucestershire Badger Group, 3 Mill Farm Drive, Paganhill, Stroud, Gloucester, GL5 4JZ.
Tel: 01453 750164

Badger watches from April to September.

Badger Breaks, Bear of Rodborough Hotel, Rodborough Common, Stroud, GL5 5DE.
Tel: 01453 87852

Organised two-day Badger watches.

Hertfordshire

Ron Denton, Herts and Middlesex Badger Group.
Tel: 07733 051760.

Badger watches at the group's hide. Bookings start on the first day of each month on a first-come first-served basis.

Herefordshire

Shortwood Family Farm, Shortwood, Pencombe, Bromyard.
Tel: 01885 400205
Fax: 01885 400720

A working farm, with a Badger sett, where Badgers can be seen most evenings.

Isle of Wight

Old Park Hotel, St Lawrence, Ventnor
Tel: 01983 852583
www.oldpark-hotel.co.uk/

The hotel specialises in holidays for families with young children. Each night food is put out on a floodlit patio for visiting Badgers. There are Red Squirrels in the grounds.

Northumberland

Phillip Wilson, West Fleetham Nature Reserve, Berwick-upon-Tweed, Northumberland.
Tel: 0191 284 6884

Good facilities, including a carpeted hide with disabled access.

Oxfordshire

College Barn Farm, Sibford Gower, Banbury.
Tel: 01295 780147
www.badger-watch.com

Short breaks in a secluded modern mobile home.
On Oxon/Warks border.

Pembrokeshire

Cwmconnell Farm, near Moylgrove in the Pembrokeshire NP.
Further details from Clwyd Badger Group (Olwen and Roger Fitzmaurice)
Tel/fax: 01244 544823
Email: clwyd.badgergroup@virgin.net

A cottage, part of a converted dairy, set above a sunken wildlife garden with pond and lovely views of the sea from rear terrace. Badgers visit terrace.

Glanmoy Lodge Guest House, Tref-Wrgi Road, Goodwick, SA64 0JX.
Tel: 01348 874 333

Guest house with Badger watching available.

Powys

Old Gwernyfed Country Manor, Felindre, Brecon, Gilfach Nature Reserve, St Harmon.
Tel: 01497 870 3011

No organised Badger watches, but Badgers are among the many wild animals that regularly visit the manor.

Gigrin Farm, Elan Valley, Rhayader Bird Reserve.
Tel: 01597 810 243

Badger watching. Also Red Kites.

Radnorshire

Tel: 01547 520313 or 01544 350603

Badger watching by arrangement for Badger Group members only.

Rutland

Anglian Birdwatching Centre,
Egleton, Oakham LE15 8BT
Tel: 01572 770651

The evening starts with an introductory slide show and hot drink in the Birdwatching Centre. The group of up to six then goes to the hide and waits for the Badgers. There is a charge of £6 per person, and it is necessary to book well in advance.

Scotland

Strathspey Badger Hide,
23 Craigie Avenue, Boat of Garten,
Inverness-shire PH24 3BL
allanbantick@hotmail.com

Organised groups visit purpose-built Badger hide. Disabled access available by prior arrangement.

Speyside Summer Mammals
Garden Office,
Inverdruie House,
Inverdruie, Aviemore,
Inverness-shire PH22 1QH
Tel: 01479 812498
www.speysidewildlife.co.uk

Badgers, Otters and Pine Martens, as well as raptors and other wildlife.

Shropshire

Badgerwatch Holidays,
Petchfield Lodge, Elton,
Nr. Ludlow SY8 2HJ.
Tel: 01568 770775

Self-catering Badger-watching holidays. Guest accommodation includes: single room, double en-suite and twin/family room en-suite. CCTV to be installed for year-round viewing.

Frank and Isobel Jones,
Frankton House, Welsh Frankton,
Nr. Oswestry, Shropshire SY11 4PA
Tel: 01691 623422

From spring for BG members and later for other selected parties. Contact through Shropshire BG.

Somerset

Secret World Badger and Wildlife
Rescue Centre, New Road Farm,
East Huntspill, Nr. Highbridge,
Somerset.
Tel: 01278 783250
www.secretworld.co.uk

Open occasional weekends to view wildlife including Badger cubs.

Suffolk

Liz Egan, Suffolk Wildlife Trust,
Brook Farm, Nr Ipswich.
Tel: 01473 890089

Open April–September.

Yorkshire
Dalby Forest Visitor Centre
Tel: 01751 472771

Organised watches from a comfortable, weatherproof hide, ten miles (16km) from Pickering, every Wednesday in summer. Wheelchair friendly.

Throughout UK
Mammals Trust UK
Tel: 020 7498 5262
www.mtuk.org/index.php?page=watching

The Mammals Trust UK runs a series of mammal-watching events in the summer months, including Badger watches.

Please note
This list of places to watch badgers has been compiled from various sources by the Badger Trust. We have checked the information as far as possible but have not visited most of the locations. The Badger Trust cannot take responsibility for any inaccuracies and recommend that individuals make their own enquiries. We would be pleased to receive any comments on the entries in this list.

There are a number of things to remember in order to watch Badgers successfully, whilst causing minimal disturbance to the animals themselves:

- Ensure that you have the permission of the landowner before you enter their land.
- Wear suitable clothing. Dark, inconspicuous clothes are best, but it is also important that clothes are warm and waterproof, since you may have to sit still for some time before you see any Badgers.
- Remain downwind when approaching the sett. Badgers have an excellent sense of smell and can find human scent particularly disturbing. Badgers often bolt underground at the slightest hint of unusual smells.
- Be as quiet as possible both when approaching and leaving the sett. Try to resist the temptation to whisper, no matter how excited you may become!
- Use insect repellent as sitting still, outside, at dawn or dusk, makes it likely that you will attract biting insects. Be warned and go prepared.

For more information on how to watch badgers, visit Steven Jackson's Brockwatch website at: www.badgers.org.uk

For more information on badgers and the work of the Badger Trust, please contact:
Susan Symes, The Badger Trust, 2b Inworth Street, London, SW11 3EP
Tel: 020 7228 6444; fax: 020 7228 6555; or email: enquiries@badgertrust.org.uk

The Mammal Watcher's Year

January/February

Remember to submit your mammal records for the previous year to your local mammal recorder. Although the short days of midwinter may not entice many people out into the cold this period can provide some exciting mammal watching as the lack of cover makes many species easier to see than at other times of the year. Hard weather often forces animals out into the open and/or into gardens to feed. Bird tables often attract mice and voles.

Female Grey Seals come ashore to moult making the species easy to see well and Common Dolphins become abundant in the seas off south-west Britain and can often be seen from the coast as they follow prey inshore. Chinese Water Deer are nearing the end of their rutting season, but the bucks remain very vocal and conspicuous during the first half of January, and Mountain Hares sporting their white winter attire can be very visible on the often snow-less hillsides. February and March is the main mating season for Badgers, and there is a lot of territorial activity with an associated rise in road casualties.

March

Although many species become more active as temperatures rise, periods of cold weather can put mammal activity and the emergence of hibernating species on hold. Certain beaches can get busy during the month as male Grey Seals join the females as they continue their moult.

Brown Hares are in the middle of their breeding season and they can provide some entertaining viewing as females repeatedly 'box' away the over-attentive males. Spring peaks of Harbour Porpoises occur off many headlands during March and April.

April

Badger activity increases towards the end of the month as young emerge from the sett for the first time. Hedgehogs emerge from hibernation. Bats continue slowly to foray from their hibernation roosts although numbers and species diversity will depend very much on local weather conditions. From late in the month family parties of Water Shrews appear and from now until mid-June is the best chance to see this species.

May

The best month to see Moles above the ground as juveniles disperse following the short breeding season, while Hedgehogs start mating and can often be located by the snuffling and snorting which is characteristic of their noisy courtship. Red Fox cubs start to emerge after several weeks in their dens, and family parties of Weasels and Stoats can occasionally be seen crossing the road at this time of year. Warm spring evenings can often be good for seeing Otters on inland waters, and female bats gather in maternity roosts.

June

A wide range of cetaceans appear off our coasts, and it is a particularly good time to look for Risso's Dolphins in western Britain. Minke Whales appear in the seas off western Scotland while Killer Whales appear in Shetland and often linger around seal colonies for days at a time. The majority of Grey Seals will now have left the beaches and sand banks to be replaced by Common Seals as the latter species' breeding season begins.

Summer is a great time to watch Badgers as the long hours of daylight mean that they often appear at the entrance of the sett before dark, while Pine Martens are easier to see well as females are often out in daylight searching for food to provide for their hungry litters.

July

Probably the best month to look for Water Voles as research has shown them to spend a good deal of their time out of the nest during this month. In Scilly, Lesser White-toothed Shrews are easier to see as they become more diurnal during summer. Edible Dormice are best searched for on warm, still evenings although it may still take some time to track down calling individuals. Bat numbers increase as many juvenile bats take their inaugural flights. Street lamps and other lights around houses can attract insects, which in turn attract bats.

Roe Deer begin rutting. At sea Atlantic White-sided Dolphins are often sighted from ferry crossings in the northern isles, and White-beaked Dolphin reports also become more frequent, particularly in Scotland. The first Fin Whales start to appear off southern Ireland, and Harbour Porpoise numbers start to build towards their autumn peak, particularly off western coasts.

August

Another excellent month for cetaceans particularly off western coasts and the Scottish islands. From mid-month, Common Seals come ashore to begin their moult. The hills and woods also become a focus of attention during August as both Red Deer and Sika Deer commence their annual rut.

This month also sees the start of the harvest, which is a very good time to observe an array of rodents as they feast on the abundance of food away from the protection of cover. Their main mammalian predators, Stoats and Weasels, also become easier to see. Young American Mink become independent during high summer and their lack of fear of humans means it is an excellent time of year to study this normally elusive animal. Many bats begin to swarm around hibernation sites.

September

Grey Seals join their Common Seal cousins on shore in September as they both start breeding from around mid-month. This is also one of the best months to observe Minke Whales off western Scotland as they feed furiously in preparation for the return migration south. Killer Whales will leave many coastlines before the end of the month while, conversely, in southern Ireland, numbers of Fin Whales continue to steadily increase.

Edible Dormice become increasingly noisy as they feed up prior to hibernating, and September is a very good month to see them. Young Polecats become independent and are often found dead on busy roads.

October

Minke Whales begin their journey south and this is a good time to spot them from headlands in the south-west, particularly from the Penwith Peninsula in Cornwall. The autumn peak of Harbour Porpoises reaches its climax in October or November with huge numbers being noted off headlands in the west of the country, especially Strumble Head in Pembrokeshire. October also sees the conclusion of the rutting season for Red, Sika and Fallow Deer.

November

Wild Cats are particularly active at this time of year and this is the best time to see them hunting during daylight. Grey Seals are still obvious on the beaches as they suckle their young, while in certain areas of the country the air reverberates to the sound of rutting Chinese Water Deer. Polecats move into farmyards from the surrounding countryside, to feed on the large numbers of rodents present.

December

Many species become increasingly elusive as winter takes a hold and several species hibernate. Fin Whales should still be present off southern Ireland, often alongside Humpback Whales, while in south-west England large schools of Common Dolphins often become visible from Cornish headlands. Winter is also a good time to watch Red Squirrels with their activity peaking in the late morning. It is also a good idea to check nest-boxes in winter, as they will often contain voles and mice, including the elusive Yellow-necked Mouse.

Useful Addresses

Please note that many of the organisations listed below are very large institutions or collaborative initiatives so the contact details vary. For specific enquiries relating to the large organisations you will need to navigate through the appropriate website.

Mammals Trust UK

www.mtuk.org
enquiries@mtuk.org

Mammals Trust UK
15 Cloisters House
8 Battersea Park Road
London
SW8 4BG

Tel: 020 7498 5262
Fax: 020 7498 4459

Many British mammals have suffered dramatic declines in numbers, particularly in the last century. For some, like the Water Vole and Red Squirrel, time is running out. For other species the threats may not be so immediate but are potentially just as serious. In response, Mammals Trust UK was set up to highlight and deal with this critical situation. It is the only charity solely dedicated to raising funds to help conserve all our native species and with your help can make a real difference.

Tracking Mammals Partnership

www.trackingmammals.org

The Tracking Mammals Partnership is a collaborative initiative, involving 24 organisations with a variety of interests in UK mammals, which aims to improve the quality, quantity and dissemination of information on the status of mammal species in the UK.

The Mammal Society

www.mammal.org.uk
enquiries@mammal.org.uk

2b Inworth Street
London
SW11 3EP

Tel: 020 7350 2200
Fax: 020 7530 2211

The Mammal Society is the only organisation solely dedicated to the study and conservation of all British mammals. For over 50 years it has been providing conservation professionals and volunteers with the skills and information needed to conserve mammals through its courses and publications. It encourages its members and local mammal groups to take part in crucial national surveys to monitor mammals; it works to protect mammals and halt the decline of threatened species. By joining The Mammal Society you, too, can help with this vital work.

People's Trust for Endangered Species

www.ptes.org
enquiries@ptes.org

People's Trust for Endangered Species
15 Cloisters House
8 Battersea Park Road
London
SW8 4BG

Tel: 020 7498 4533
Fax: 020 7498 4459

Since 1977 PTES has been helping to ensure a future for many endangered species throughout the world. People are becoming increasingly aware of the threats to wildlife and the alarming rate at which the numbers of many species are declining. The society is committed to working to preserve them in their natural habitats for future generations to enjoy. They focus on specific problems as they arise and regularly write to their supporters outlining each new project they undertake.

The Badger Trust

www.badger.org.uk
enquiries@badgertrust.org.uk

2b Inworth Street
London
SW11 3EP

Tel: 020 7228 6444
Fax: 020 7228 6555

The Badger Trust is a registered charity that promotes the conservation, welfare and protection of Badgers, their setts and habitats. It is the leading voice for Badgers in Britain and represents and supports 80 local voluntary Badger groups.

The Whale and Dolphin Conservation Society

www.wdcs.org
info@wdcs.org

WDCS
Brookfield House
38 St Paul Street
Chippenham
Wiltshire SN15 1LY

Tel: 0870 870 0027
Fax: 0870 870 0028

WDCS is dedicated to the conservation and welfare of all whales, dolphins and porpoises.

The Bat Conservation Trust

www.bats.org.uk
enquiries@bats.org.uk

The Bat Conservation Trust
Unit 2, 15 Cloisters House
8 Battersea Park Road
London
SW8 4BG

Tel: 020 7627 2629
Fax: 020 7627 2628

The Bat Conservation Trust (BCT) works to ensure that bat populations survive for future generations to enjoy. It is the only national organisation solely devoted to bat conservation, and acts as the voice for bats in the UK.

The Royal Society for the Protection of Birds (RSPB)

www.rspb.org.uk

The Lodge
Sandy
Bedfordshire
SG19 2DL

Tel: 01767 680551

The RSPB is one of Britain and Ireland's most influential nature conservation organisations, boasting over one million members. The organisation owns or manages 200 nature reserves throughout the UK, covering an area in excess of 100,000 hectares.

The National Trust

www.nationaltrust.org.uk
enquiries@thenationaltrust.org.uk

The National Trust acts as a guardian for the nation in the acquisition and protection of threatened coastline, countryside and buildings.

The Forestry Commission

www.forestry.gov.uk
enquiries@forestry.gsi.gov.uk

The Forestry Commission is the Government Department responsible for forestry policy throughout Great Britain.

Natural England (formerly English Nature)

www.naturalengland.org.uk

Natural England is the Government-funded body whose purpose is to promote the conservation of England's wildlife and natural features. It works closely with the Joint Nature Conservation Committee, Scottish Natural Heritage and the Countryside Council for Wales for a consistent approach to nature conservation throughout Great Britain, and towards fulfilling international obligations.

The Wildlife Trusts

www.wildlifetrusts.org
enquiry@wildlife-trusts.cix.co.uk

The Wildlife Trusts partnership is the UK's leading conservation charity exclusively dedicated to wildlife. The Wildlife Trusts manage over 2,560 nature reserves, ranging from rugged coastline to urban wildlife havens. With more than 560,000 members, and unparalleled grass-roots expertise, The Wildlife Trusts lobby for better protection of the UK's natural heritage and are dedicated to protecting wildlife for the future.

The Wildfowl and Wetlands Trust

www.wwt.org.uk
info.slimbridge@wwt.org.uk

The Wildfowl & Wetlands Trust is the largest international wetland conservation charity in the UK. WWT's mission is to conserve wetlands and their biodiversity. These are vitally important for the quality and maintenance of all life.

Other useful websites

British Deer Society	www.bds.org.uk
Countryside Council for Wales	www.ccw.gov.uk
Deer Commission for Scotland	www.dcs.gov.uk
Joint Nature Conservation Committee	www.jncc.gov.uk
Scottish Natural Heritage	www.snh.gov.uk
Wildlife Conservation Research Unit	www.wildcru.org
World Wide Fund For Nature (WWF)	www.panda.org.uk
	www.wwf-uk.org

Whale-watching companies

Such is the current popularity of cetacean watching that many tour companies have sprung up that take visitors out to sea in productive areas. For a full range of whale- and dolphin-watching companies operating throughout the United Kingdom and Republic of Ireland visit the Whale and Dolphin Conservation Society website: www.wdcs.org, and then follow links to the whale-watching database.

Companies offering trips along the west coast of Scotland and particularly around the Inner Hebrides include:

- MV Amidas: www.mv-amidas.com
- Ardnamurchan Charters: www.west-scotland-marine.com
- Guideliner Charters: Week-long wildlife trips through the Hebrides: www.guideliner.co.uk
- Hebridean Adventure: Wildlife trips onboard classic sailing vessel: www.hebrideanadventure.co.uk
- Inter-Island Cruises: Daily whale watching and island trips from Croig, Mull: www.whalewatchingtrips.co.uk
- Lady Jayne-Colonsay: Daily boat trips leaving from Colonsay: www.colonsay.org.uk/boat/1boat.html
- MV Cuma: Liveaboard wildlife cruises from the Western Isles: www.island-cruising.com
- MV Volante: Daily wildlife, angling and sightseeing trips: www.volanteiona.com/index.html

- Northern Light Charters: Dive and wildlife charters from Oban: www.northernlight-uk.com
- Staffa Boat Trips: Daily boat trips to Staffa Island: www.staffatrips.f9.co.uk
- Turus Mara: Daily boat trips to Staffa and the Treshnish Isles: www.turusmara.com
- Sail Gairloch: www.porpoise-gairloch.co.uk
- Sea Life Surveys: Whale, dolphin and birdwatching packages throughout the Hebrides: www.sealifesurveys.com
- Sealife Adventures: Whale watching and wildlife in the Firth of Lorn: www.sealife-adventures.com

Companies offering trips to see the resident population of Bottle-nosed Dolphins of the Moray and Cromarty Firths include:

- Ecoventures: www.ecoventures.co.uk
- Inverness Dolphin Cruises: www.inverness-dolphin-cruises.co.uk/map.html

Tour Companies

In addition to specialised whale-watching trips, several organisations offer excursions and holidays that are specifically designed to see mammals. These include:

- Aigas Field Centre: www.aigas.co.uk
- Great Glen Wildlife: www.greatglenwildlife.co.uk
- Speyside Wildlife: www.speysidewildlife.co.uk

County Recorders
England & The Isle of Man

Bedfordshire
Michael McCarrick, 'Mammal Recorder', Bedfordshire Natural History Society, 38 Duncombe Close, Luton, Bedfordshire, LU3 2HR
mick.mccarrick@btinternet.com - Preferred Method of Contact

Berkshire
Adrian Hutchings, Berkshire Manager, Thames Valley Environmental Records Centre, Planning, Council Offices, Market Street, Newbury, Berkshire, RG14 5LD
01635 519179
ahutchings@westberks.gov.uk

Bristol
(including south Gloucester, north & northeast Somerset and Bath)
David Trump, Bristol Mammal Group, Windrush, West End Lane, Nailsea, North Somerset, BS48 4DB
d.m.trump@tinyworld.co.uk - Preferred Method of Contact

Buckinghamshire
Martin Harvey, Environmental Records Officer, Buckinghamshire & Milton Keynes Environmental Records Centre, Museum Resource Centre, Tring Road, Halton, Aylesbury, Buckinghamshire HP22 5PN
01296 696012
01296 624519 (Fax)
mcharvey@buckscc.gov.uk

Cambridgeshire
Howard Hillier, Cambridgeshire Mammal Group, 127 Fletton Avenue, Peterborough, Cambridgeshire, PE2 8BX

Cheshire
rECOrd, Chester Zoological Gardens, Upton, Chester, Cheshire, CH2 1LH
01244 383749
01244 383569 (Fax)
info@rECOrd-LRC.co.uk

Cornwall
Miss Alex Howie, Mammals Project Co-ordinator, ERCCIS, Cornwall Wildlife Trust, Five Acres, Allet, Truro, Cornwall, TR4 9DJ
alex@cornwt.demon.co.uk - Preferred Method of Contact

Cumbria
Stephen Hewitt, Tullie House Museum & Gallery, Castle Street, Carlisle, Cumbria, CA3 8TP
steveh@carlisle.gov.uk - Preferred Method of Contact

Derbyshire
Nick Moyes, Assistant Keeper of Natural History & Records, Derbyshire Museum and Art Gallery, The Strand, Derby, DE1 1BS
01332 716655

Derek Whiteley, Recorder, Derbyshire Mammal Group, Beech Cottage, Wardlow, Derbyshire, SK17 8RP
derek@thedeadtree.wanadoo.co.uk - Preferred Method of Contact

Devon
Eleanor Bremner, Devon Biodiversity Record Centre, Shirehampton House, 35-37 St David's Hill, Exeter, Devon, EX4 4DA
01392 279244
01392 433221 (Fax)
ebremner@devonwt.cix.co.uk

Dorset
John Stobart, 1 Grange Cottages, Chetnole, Sherborne, Dorset, DT9 6PE
john@country-side.freeserve.co.uk - Preferred Method of Contact

Durham
Kevin O'Hara, Northumberland Wildlife Trust, The Garden House, St Nicholas Park, Jubilee Road, Newcastle-upon-Tyne, NE3 3XT
0191 584 3112
kevin.ohara@northwt.org.uk

Essex
John Dobson, 148 Main Road, Danbury, Essex, CM3 4DT
johndobson@mammals.fsnet.co.uk - Preferred Method of Contact

Gloucestershire
Rosie Cliffe, Gloucestershire Wildlife Trust, Dulverton Building, Robinswood Hill Country Park, Reservoir Road, Gloucester, Gloucestershire, GL6 4SX
01452 383333,
01452 383334 (Fax)
RosieC@gloucswt.cix.co.uk

Greater London
(Including southern Hertfordshire and northern Surrey)
Clive Herbert, County Recorder, London Natural History Society, 67a Ridgeway Avenue, East Barnet, Hertfordshire, EN4 8TL
020 8440 6314 (Home)
020 8440 6314 (Work)
020 8440 6314 (Fax)

Greater Manchester, Lancashire, Merseyside
Dr Clem Fisher, Liverpool Museum, William Brown Street, Liverpool, L3 8EN
0151 207 0001
0151 922 7845
clemf@nmgmzoo2.demon.co.uk

Hampshire
Dr Sarah Benge, Hampshire Mammal Recorder, Hampshire Mammal Group, 1 Southside Cottages, Longstock, Stockbridge, Hampshire, SO20 6DN
seb@soton.ac.uk - Preferred Method of Contact

Herefordshire
Hilary Smith, 89 College Road, Hereford, HR1 1ED
01432 357732
herefordwt@cix.co.uk

Hertfordshire
Ms Jenny Jones, 23 North Road, Hertford, SG14 1LN,
01992 581442
jenny.jones@hertscc.gov.uk

Huntingdon
Henry Arnold, Centre for Ecology and Hydrology, Monks Wood, Abbots Ripton, Huntingdon, Cambridgeshire, PE28 2LS
01487 772406

Isle of Wight
Richard Grogan, IOW Natural History & Archaeological Society, The Carriage House, Station Road, Ningwood, Newport, Isle of Wight, PO40 4NJ
01983 872 693

Kent
Peter Heathcote, 9 Greenfinches, New Barn, Longfield, Kent, DA3 7ND
01474 704298,
peter@heathcote100.freeserve.co.uk

Leicestershire
Darwyn Sumner, Holly Hayes Environmental Records Centre, 216 Birstall Road, Birstall, Leicestershire, LE4 4DG
0116 267 1950,
dsumner@leics.gov.uk

Lincolnshire
Colin Faulkner, 65 London Road, Spalding, Lincolnshire PE11 2TH
01775 766286
a.faulkner@care4free.net

Norfolk
Mike Toms, GBW Organiser, British Trust for Ornithology, The Nunnery, Thetford, Norfolk, IP24 2PU
01842 750050
01842 750030 (Fax)
michael.toms@bto.org

Northamptonshire
Phil Richardson, 10 Bedford Cottages, Great Brington, Northamptonshire, NN7 4JE
Prichabat@aol.com - Preferred Method of Contact

Northumberland
Kevin O'Hara, Northumberland Wildlife Trust, The Garden House, St Nicholas Park, Jubilee Road, Newcastle-upon-Tyne, NE3 3XT
0191 584 3112
kevin.ohara@northwt.org.uk

Nottinghamshire
John Ellis, Records and Information Officer, Nottinghamshire Wildlife Trust, The Old Ragged School, Brook Street, Nottingham, Nottinghamshire, NG1 1EA
jellis@nottswt.cix.co.uk - Preferred Method of Contact

Oxfordshire
Gavin Bird, Oxfordshire Records Centre Manager, Thames Valley Environmental Records Centre, Fletchers House, Park Street, Woodstock, Oxfordshire, OX20 1SN
01993 814147
07776 461279 (Mobile)
01993 814116 (Fax)
gavin.bird@oxfordshire.gov.uk

Shropshire
Dr John Mackintosh, Lawton Cottage, Stanton Lacey, Ludlow, Shropshire, SY8 2AL
01584 861688
john.mackintosh@ntlworld.com

Somerset
Somerset Environmental Record Centre, Tonedale Mill, Wellington, Somerset TA21 0AW
01823 664450
01823 652411 (Fax)
info@somerc.com

Sorby, (Sheffield and North Derbyshire)
Valerie Clinging, Recorder, Sorby Natural History Society, 44 Alms Hill Road, Sheffield, Yorkshire, S11 9RS
0114 236 7028
v.clinging@sheffield.ac.uk - Preferred Method of Contact

Staffordshire
Derek Crawley, County Recorder - Staffordshire, 155a Eccleshall Road, Stafford, Staffordshire, ST16 1PD
01785 258993 (Home)
01785 712209 (Work)
07904 026507 (Mobile)
01785 715701 (Fax)
derek.crawley@rodbaston.ac.uk

Suffolk
Dr Simone Bullion, Suffolk Wildlife Trust, Brooke House, The Green, Ashbocking, Ipswich, Suffolk, IP6 9JY
01473 890089
01473 890165 (Fax)
simoneb@suffolkwildlife.cix.co.uk

Surrey
Dave Williams, Mammal Officer, Surrey Wildlife Trust, School Lane, Pirbright, Woking, Surrey, GU24 0J
01483 810214 (Home)

01483 795454 (Work)
07768 518064 (Mobile)
01483 486505 (Fax)
dave.williams@surreywt.org.uk

Sussex

Penny Green, Sussex Biodiversity Records Centre, Woods Mill, Henfield, West Sussex, BN5 9SD
01273 497521
01273 494500 (Fax)
pennygreen@sussexwt.org.uk

Tees Valley (Unitary Authority of Stockton-on-Tees, Hartlepool, Middlesborough and Redcar & Cleveland)

Jeremy Garside, Conservation Officer, Tees Valley Wildlife Trust, Bellamy Pavilion, Kirkleatham Old Hall, Redcar, Cleveland, TS10 5NW

Warwickshire

Tony Ware, Warwickshire Mammal Group, Brandon Marsh Nature Centre, Brandon Lane, Coventry, Warwickshire, CV3 3GW
0121 680 7723

West Midlands

Tim Moughtin, 4 Rosliston Road, Walton upon Trent, Swadlincote, Derbyshire, DE12 8NQ
01283 713498 (Home)
0121 454 1199 (Work)
0121 454 6556 (Fax)
timmoughtin@hotmail.co.uk

Wiltshire

Mark Satinet, Wiltshire Wildlife Trust, Elm Tree Court, Long Street, Devizes, Wiltshire, SN10 1NJ
07747 678390 (Mobile)
mark.satinet@hyderconsulting.com
Preferred Method of Contact - Work email or mobile

Worcestershire

Shaun Micklewright, 45 Manor Avenue South, Kidderminster, Worcestershire, DY11 6DE
01562 824687
shaun.micklewright@peoplepc.co.uk

Yorkshire

Colin Howes, Mammal Recorder, Yorkshire Naturalists Union, Museum & Art Gallery, Chequer Road, Doncaster, South Yorks, DN1 2AE
01302 734287
01302 735409 (Fax)
colin.howes@doncaster.gov.uk

Isle of Man

Ed Pooley, Ballasoalt, Colby, Isle of Man, IM9 4HN
01624 834739
ej@ballasoalt.com, Isle of Man. Contact by email only.

Northern Ireland

Damian McFerran, Records Centre Manager, Centre for Environmental Data and Recording (CEDaR), Ulster Museum Botanic Gardens, Belfast, BT9 5AB
01232 383154
01232 383103 (Fax)
or
Mrs Lynne Rendle, Vertebrate Recorder, Centre for Environmental Data and Recording (CEDaR), Botanic Gardens, Ulster Museum Belfast, BT9 5AB
lynne.rendle@magni.org.uk - Preferred Method of Contact

Scotland

Aberdeenshire
David Simmons, 2 The Village Square, Monymusk, Aberdeenshire, AB51 8HJ
01467 651586

Angus and Dundee
Richard, Brinklow, Dundee Museum Local Records Centre, Natural History Museum, Dundee, Angus & Forfar, DD1 4PG
01382 432067

Argyll and Inner Hebrides
Situation Vacant – Please contact The Mammal Society if you can take on this role.

Ayrshire
Situation Vacant – Please contact The Mammal Society if you can take on this role.

Borders
Dr Jon Mercer, Scottish Borders Biological Records Centre, Harestanes Countryside Visitor Centre, Ancrum, Jedburgh, Roxburghshire, TD8 6UQ
01835 830405
01835 830734 (Fax)
SBBRC@scotborders.gov.uk

Central
John Haddow, 27 Balmoral Court, Dunblane, Perthshire, FK15 9HQ
01786 823390
07801 441288 (Mobile)
john.haddow@virgin.net

Dumfries and Galloway
Stuart Spray, 43 Sydney Place, Lockerbie, Dumfriesshire, DG11 2JB
07810 814554 (Mobile)
stuart@spray.idps.co.uk

East Lothian
Situation Vacant – Please contact The Mammal Society if you can take on this role.

Fife
Dr Gordon Corbet, Little Dumbarnie, Upper Largo, Leven, Fife, KY8 6JG
01333 340634
gcorbet@dumbarnie.freeserve.co.uk
Short reports by email, longer by post

Highland
Ms Ro Scott, Highland Biological Recording Centre, Peddieston Cottage, Farness, Cromarty, Ross-shire, IV11 8XX
01381 600392
ro.scott@care4free.net

Orkney
Martin Gray, Rue, North Ronaldsay, Orkney

Outer Hebrides
Bill Neill, Rannachan, Askernish, South Uist, Western Isles, HS8 55Y
01878 700237

Perth and Kinross
Mark Simmons, Principal Officer Natural Sciences, Perth Museum and Art Gallery, George Street, Perth, PH1 5LB
mjsimmons@pkc.gov.uk - Preferred Method of Contact

Shetland
Paul Harvey, Shetland Biological Records Centre, Shetland Amenity Trust, Garthspool, Lerwick, Shetland, ZE1 0NY
01595 694688 (Work)
07818 632413 (Mobile)
01595 693956 (Fax)
sbrc@zetnet.co.uk

Skye, Loch Alsh, The Small Isles & The Islands of Islay, Raasay, and Scalpay
Paul Yoxon, Isle of Skye Environmental Centre Ltd, Broadford, Isle of Skye, IV49 9AQ
01471 822487

Strathclyde and the Clyde Faunal Area
(Ayrshire; Renfrewshire; Lanarkshire; Dunbartonshire; West Stirlingshire; Loch Lomond; South Argyll; Buteshire; Ailsa Craig & the small Clyde Islands; Arran; Bute and Cumbrae)
Dr J.A. Gibson, Scottish Natural History Library, Foremount House, Kilbarchan, Renfrewshire, PA10 2EZ
01505 702419

Wales

Anglesey and Merioneth
Ms Jean Matthews, County Mammal Recorder, c/o Countryside Council for Plas Penrhos Campus, Penrhos Road, Bangor, Gwynedd, LL57 2BQ
j.matthews@ccw.gov.uk - Preferred Method of Contact

Breconshire
Phil Morgan, Llys Newydd, Llanfihangel Talyllyn, Brecon, Powys, LD3 7TG
01874 658650
07713 258937 (Fax)
phill@justmammals.co.uk

Caernarfonshire
Peter Wells, Flat 2, Bryn Tirion Mawr, Bryn Tirion Mawr, Abergwyngregyn, Llanfairfechan, Caernarfonshire/Sir Gaernarfon, LL33 0LE
01248 680818

Carmarthenshire
Neil Matthew, Carmarthen Bat Group, Lamb Shop, Kilycwm, Llandovery, Carmarthenshire, SA20 0SS
01550 721542

Ceredigion
Tom McOwat, 19 Parc Puw, Drefach, Velindre, Llandysul, Ceredigion, SA44 5UZ
01454 572112

Denbighshire
Brian Burnett, Nant Yr Hafod Cottage, Llandegla, Wrexham, Denbigh, LL11 3BG
indesigneko@aol.com

Flint and Wrexham
Mike Griffiths, 'Preswylfa', 1 Heol Llandewi, Wrexham, Denbigh, LL12 8JR
01978 264516

Glamorgan
Dr Dan Forman, Department of Biological Sciences, Institute of Environment and Society, University of Swansea, Singleton Park, Swansea, Vale of Glamorgan, SA2 8PP
d.w.forman@swansea.ac.uk - Preferred Method of Contact

Monmouthshire
Dr Peter Smith, Principal Ecologist, Smith Ecology Ltd, 1 Bettws Cottage, Bettws, Abergavenny, Monmouthshire, NP7 7LG
01873 890598 (Home)
01873 890055 (Work)
01873 890976 (Fax)
peter@smithecology.co.uk

Montgomeryshire
Andrew McLeish, Alana Ecology, The Old Primary School, Church Street, Bishops Castle, Shropshire, SY9 5AE
01686 670643 (Home)
01588 630173 (Work)
01588 630176 (Fax)
amcleish@alanaecology.com

Pembrokeshire
Annie Haycock, 1 Rushmoor, Martletwy, Pembroke, SA67 8BB
rushmoor1@tiscali.co.uk

Radnorshire
Julian Jones, Conservation Manager, Radnorshire Wildlife Trust, Warwick House, High Street, Llandrindod Wells, Radnorshire, LD1 6AG
01597 823298
01597 823298 (Fax)
info@radnorshirewildlifetrust.org.uk

Local Mammal Groups

For further information please refer to The Mammal Society website: www.abdn.ac.uk/mammal/index.shtml

Birmingham and the Black Country Mammal Group

Sara Carvalho
The Wildlife Trust for Birmingham
& the Black Country
28 Harborne Road
Edgbaston
Birmingham B15 3AA
enquiries@ecorecord.org.uk

Brecknock Mammal Group

Phil Morgan
Brecknock Wildlife Trust
Lion House
Bethal Square
Brecon LD3 9EH

Bristol Mammal Group

David Trump
Windrush
West End Lane
Nailsea
North Somerset BS48 4DB
d.m.trump@tinyworld.co.uk

Cambridgeshire Badger and Otter Group

Peter Pilbeam
6 Cross Keys Court
Cottenham
Cambridge CB4 8UW
pgp1@admin.cam.ac.uk

Cambridgeshire Mammal Group

Howard Hillier
17 St Margarets Road
Old Fletton
Peterborough PE2 9EA

Cheshire Mammal Group

Tony Parker
41 The Park
Penketh
Warrington
Cheshire WA5 2SG
tony.parker01@btinternet.com

Clyde Mammal Group

Dr J.A. Gibson
Scottish Natural History Library
Foremont House
Kilbarchan
Renfrewshire PA10 2EZ

Cornwall Mammal Group

Ian Bennallick
Cornwall Wildlife Trust
Five Acres, Allet
Truro
Cornwall TR4 9DJ

ian@cornwt.demon.co.uk

Cumbria Mammal Group

Joe Murphy
c/o Cumbria Wildlife Trust
Plumgarths
Crook Road Kendal
Cumbria LA8 8LX

Derbyshire Mammal Group

Chair: Dave Mallon
d.mallon@zoo.co.uk
Secretary: Anna Evans

www.derbyshiremammalgroup.com

benanna@adventures.fsworld.co.uk

Devon Mammal Group

Antony Bellamy
58 Oxford Street
St Thomas
Exeter EX2 9AG

www.devonmammalgroup.org
enquiries@devonmammalgroup.org

Essex Mammal Group

Louise Wells
113 Westwood Road
Seven Kings
Ilford
Essex

lwells@wildlondon.org.uk

Gloucestershire Mammal Group

Rosie Cliffe
Gloucestershire Wildlife Trust
Dulverton Buildings
Robinswood Hill Country Park
Reservoir Road
Gloucester GL4 9SX

Hampshire Mammal Group

Kirsten Knapp
Beechcroft House
Vicarage Lane
Eastleigh SO32 2DP

Herefordshire Action for Mammals

Hilary Smith
89 College Road
Hereford HR1 1ED

wildways.hilary@bigfoot.com

Hertfordshire Mammal Group

Peter Oakenfull poakenfull@aol.com
16 Little Lake
Welwyn Garden City AL7 4RT

Kent Mammal Group

Ken West
Beach Cottage,
Sea Wall,
Whitstable,
Kent CT5 1BX

Leicestershire Mammals

Darwyn Sumner dsumner@leics.gov.uk
Holly Hayes Environmental Records Centre
216 Birstall Road
Birstall
Leicestershire LE4 4DG

Northern Ireland Mammal Group

Dr Kate O'Neill, Quercus, k.oneill@qub.ac.uk
School of Biology & Biochemistry,
MBC, 97 Lisburn Road,
Belfast BT9 7BL

Northumbria Mammal Group (Northumberland, Durham, Tyneside, Teeside)

Veronica Carnell veronica_carnell@hotmail.com
44 Prince's Meadow
Newcastle upon Tyne NE3 4RZ

Somerset Mammal Group

Peter Allwright
Silver Birches, Spring Lane
Moorlinch
Bridgwater
Somerset TA7 9DD

Sorby Mammal Group

Derek Whiteley
6 Pancake Row
Cressbrook
Derbyshire SK17 8SY

South Essex Action for Mammals

Mike O'Connor mike@oconnor1357.freeserve.co.uk
3 Priorywood Drive
Leigh on Sea
Essex SS9 4BU

Snowdonia Mammal Group

Kate Williamson kate.williamson@eryri-npa.gov.uk
Snowdonia National Park Authority,
National Park Office
Penrhyndeudraeth
Gwynedd LL48 6LF

Staffordshire Mammal Group

Derek Crawley
155a Eccleshall Road
Stafford
Staffordshire ST16 1PD

Surrey Mammal Group

Dave Williams
Surrey Wildlife Trust
School Lane
Pirbright
Surrey GU24 0JN

Sussex Mammal Recording Group

Neil Mitchell sxmrg@hotmail.com
15 Woodside
Barnham
Bognor Regis
West Sussex PO22 0HZ

Warwickshire Mammal Group

Tony Ware
Brandon Marsh Nature Centre
Brandon Lane
Coventry CV3 3GW

Working for Mammals Group (Hampshire)

Jim Park
61 Park Road
Alverstoke
Gosport
Hampshire PO12 2HQ

Yorkshire Mammal Group

Denise Ray ymg@netcomuk.co.uk
Barn House, North Farm
Scoreby
York YO41 1NP

Local Bat Groups

A network of over 90 volunteer-based bat groups and bat workers, all working to protect and conserve the country's bats, covers the UK. With around 4,000 members, these groups are the key to bat conservation in the UK. They carry out much of the practical work involved in monitoring bat populations and their habitats across the country.
Bat groups marked * do not have a website. In these cases please contact the Bat Conservation Trust helpline on 0845 1300 228, who can put you in contact with your local group.

Southeast England
Hertfordshire and Middlesex Bat Group: www.hertsmiddlesexbatgroup.org.uk
Berkshire and South Buckinghamshire Bat Group: www.berksbats.org.uk
Bedfordshire Bat Group: www.batsinbeds.org
North Buckinghamshire Bat Group: www.northbucksbatgroup.org.uk
Hampshire Bat Group: http://homepage.ntlworld.com/glio/index1.htm
Isle of Wight Bat Group: http://mysite.wanadoo-members.co.uk/iow-bathospital
Oxfordshire Bat Group*
Wiltshire Bat Group: contact: shlaurence@aol.com
Five Rivers Bat Association: contact: Phil Smith, 63 Shaftesbury Road, Wilton, Salisbury, Wiltshire, SP2 0DU.
London Bat Group: www.londonbats.org.uk
Kent Bat Group: www.kentbatgroup.org.uk
Surrey Bat Group: www.surreybats.org.uk
Sussex Bat Group: www.sussexbatgroup.org.uk

Southwest England
Avon Bat Group: www.avonbatgroup.co.uk
Cornwall Bat Group: www.cornwall-batgroup.co.uk
Devon Bat Group: www.dbg.me.uk
Dorset Bat Group: www.dorsetbatgroup.org.uk
Somerset Bat Group: contact Dinah Harding, Tisledown, Ilton, Ilminster, Somerset, TA19 9HL.

Channel Islands
Alderney*
Guernsey: contact Pat Costen, La Broderie, La Claire Mare, St.Peter's Port, Guernsey, GY7 9QA. Email: pcosten@guernsey.net
Jersey*

East Anglia
Cambridgeshire Bat Group: www.cambsbats.co.uk
Essex Bat Group*
Norfolk Bat Group: www.norfolk-bat-group.org.uk
Suffolk Bat Group*

Central England
Herefordshire Bat Group: www.fly.to/hereford_bats
Leicestershire and Rutland Bat Group*
Northamptonshire Bat Group*
South Nottinghamshire Bat Group: www.southnottsbatgroup. org.uk
North Nottinghamshire Bat Group: www.nottsbats.org.uk
West Midlands Bat Groups (incorporating Birmingham & Black Country, Shropshire, South-east Staffordshire Bat Groups): www.westmidlands-batgroups.org.uk
Derbyshire Bay Conservation Group: www.derbyshirebats.org.uk
Gloucestershire Bat Group*
Staffordshire Bat Group: www.staffsbatgroup.org.uk
Warwickshire Bat Group: www.warksbats.co.uk

Northern England
Lincolnshire Bat Group*
Cheshire Bat Group: www.consult-eco.ndirect.co.uk/lrc/batg.htm
Durham Bat Group: www.durhambats.org.uk
Northumberland Bat Group: bats@ryal.freeserve.co.uk
Yorkshire (East)*
North Yorkshire Bat Group: www.nyorkbats.freeserve.co.uk
Yorkshire (Barnsley)*
Yorkshire (Sorby-Sheffield)*
Yorkshire (Wakefield)*
West Yorkshire Bat Group: http://web.ukonline.co.uk/wybg/frame.htm
Cumberland Bat Group: www.cumbriabats.org.uk
Greater Manchester*
Lancashire (East)*
North Lancashire Bat Group: www.nlbg.org.uk
South Lancashire Bat Group: www.slbg.org.uk
Isle of Man*

Wales
Brecknock Bat Group: contact Diane Morgan, c/o Brecknock Wildlife Trust, Lion House, Bethel Square, Brecon, LD3 7AY. Email: brecknockwt@cix.co.uk
Carmarthenshire*
Dyfed*
Montgomery*
North Ceredigion Bat Conservation Group: http://users.aber.ac.uk/ltt/bats/
Pembrokeshire Bat Group: www.pembsbats.org.uk
Radnorshire*
Clwyd Bat Group: contact Mike Castle. Email: mike.castle@which.net
Gwynedd*
Gwent Bat Group: www.gwentbatgroup.org.uk
Glamorgan*
Cardiff*
Welsh Bat Groups: www.welshbats.org.uk

Scotland
Aberdeen*
Orkney*

Strathspey: contact Anne Elliott: Email: anne.elliott@snh.gov.uk
Moray*
Inverness: contact Jonathan Watt. Email: jonathan.watt@highland.gov.uk
Skye: contact Grace Yoxon. Email: iosf2@aol.com
Lochaber: contact Christine Welsh. Email: christine.welsh@snh.gov.uk
North Highland Bat Network (Sutherland, Caithness and NW Ross): contacts: lazytrout@care4free.net or nredgate@ndres.co.uk
Angus: contact Tim Caselton, Ranger Centre, Monikie Country Park, Monikie, Angus, DD5 3QN. Email: tim.caselton@btinternet.com
Perth*
Fife*
Central Scotland*
Lothians*
Kinross: contact sarah.eaton@care4free.net
Borders*
Isle of Arran*
Clyde Bat Group: www.clydebatgroup.co.uk
Dumfries and Galloway*
Ayrshire: contact Louise Kirk (01294 225198) or Garry Nixon (01560 323257)
Loch Lomond*
Scottish Bats: www.scotbats.org.uk

Northern Ireland
Contact: Lynne Rendle. Email: lynne.rendle.um@nics.gov.uk

Republic of Ireland
Cork Bat Group: www.iol.ie/~corkbatgroup/index.htm
Dublin Bat Group: www.geocities.com/dublinbat

Appendix
Scientific names of other species mentioned

Mammals

Nomenclature follows *Mammals of the World: A Checklist* (Duff and Lawson 2004).

American Beaver	*Castor canadensis*
European Beaver	*Castor fiber*
Golden Hamster	*Mesocricetus auratus*
Muskrat	*Ondatra zibethicus*
Indian Crested Porcupine	*Hystrix indica*
Coypu	*Myocastor coypus*
Iriomote Cat	*Felis bengalensis iriomotensis*
Spanish Lynx	*Felis pardina*
American Marten	*Martes americana*
European Mink	*Mustela lutreola*
Northern (Steller's) Sea-Lion	*Eumetopias jubatus*
Indian Muntjac	*Muntiacus muntjac*
Père David's Deer	*Elaphurus davidianus*

Birds

Nomenclature follows the British Ornithologists' Union's *The British List* (2006)

Mute Swan	*Cygnus olor*
Common Eider	*Somateria mollissima*
Willow (Red) Grouse	*Lagopus lagopus*
Capercaillie	*Tetrao urogallus*
Manx Shearwater	*Puffinus puffinus*
(Northern) Gannet	*Morus bassanus*
(Great) Bittern	*Botaurus stellaris*
Honey Buzzard	*Pernis apivorus*
White-tailed Eagle	*Haliaeetus albicilla*
Hen Harrier	*Circus cyaneus*
Marsh Harrier	*Circus aeruginosus*
Goshawk	*Accipiter gentilis*
Common Buzzard	*Buteo buteo*
Golden Eagle	*Aquila chrysaetos*
Osprey	*Pandion haliaetus*
Avocet	*Recurvirostra avosetta*
Little Tern	*Sternula albifrons*
(Black-legged) Kittiwake	*Rissa tridactyla*
Puffin	*Fratercula arctica*
Tawny Owl	*Strix aluco*
Short-eared Owl	*Asio flammeus*
Common Swift	*Apus apus*
(Common) Nightingale	*Luscinia megarynchos*
Crested Tit	*Lophophanes cristatus*

Bearded Tit	*Panurus biarmicus*
(Common) Starling	*Sturnus vulgaris*
Parrot Crossbill	*Loxia pytyopsittacus*
Scottish Crossbill	*Loxia scotica*

Butterflies

Nomenclature follows *The Millennium Atlas of Butterflies in Britain and Ireland* (Warren *et al.* 2001).

Lulworth Skipper	*Thymelicus action*
Grizzled Skipper	*Pyrgus malvae*
Swallowtail	*Papilio machaon*
Wood White	*Leptidea sinapis*
Brown Hairstreak	*Thecla betulae*
Black Hairstreak	*Satyrium pruni*
Chalkhill Blue	*Polyommatus coridon*
Adonis Blue	*Polyommatus bellargus*
Large Blue	*Maculinea arion*
White Admiral	*Limenitis camilla*
Purple Emperor	*Apatura iris*
Small Pearl-bordered Fritillary	*Boloria selene*
Pearl-bordered Fritillary	*Boloria euphrosyne*
Heath Fritillary	*Melitaea athalia*

Dragonflies

Nomenclature follows British Dragonfly Society Checklist

Azure Hawker	*Aeshna caerulea*
Brilliant Emerald	*Somatochlora arctica*
Highland Darter	*Sympetrum nigrescens*
White-faced Darter	*Leucorrhinia dubia*

Plants

Nomenclature follows *Wild Flowers of Britain and Ireland* (Blamey *et al.* 2003)

Scots Pine	*Pinus sylvestris*
Wild Service	*Sorbus torminalis*
Sessile Oak	*Quercus petraea*
Holly	*Ilex aquifolium*
Beech	*Fagus sylvatica*
Mistletoe	*Viscum album*
Hornbeam	*Carpinus betulus*
Bilberry	*Vaccinium myrtillus*
Heather	*Calluna or Erica* spp.
Oxlip	*Primula elatior*
Honeysuckle	*Lonicera* spp.
Fritillary	*Fritillaria meleagris*
Bluebell	*Hyacinthoides non-scripta*
Wild Daffodil	*Narcissus pseudonarcissus*
Bulrush	*Typha latifolia*
Autumn Lady's Tresses	*Spiranthes spiralis*
Lady Orchid	*Orchis purpurea*
Burnt-tip Orchid	*Orchis ustulata*

Pyramidal Orchid	*Anacamptis pyramidalis*
Green-winged Orchid	*Orchis morio*
Early Marsh Orchid	*Dactylorhiza incarnata*
Common Reed	*Phragmites australis*
Canary Grass	*Phalaris canariensis*

Reptiles & Amphibians

Nomenclature follows *Reptiles and Amphibians of Britain and Europe* (Arnold & Burton 1978)

Great Crested Newt	*Triturus cristatus*
Smooth Newt	*Triturus vulgaris*
Natterjack Toad	*Bufo calamita*

Further Reading

Altringham J. 2003. *British Bats*. Collins New Naturalist series.

Battersby, J. 2005. *UK Mammals: Species Status and Population Trends*. The Tracking Mammals Partnership.

Battersby, J. (Ed.) & Tracking Mammals Partnership 2005. *UK Mammals: Species status and population trends. First Report by the Tracking Mammals Partnership*. JNCC/Tracking Mammals Partnership, Peterborough.

Birks, J.D.S. & Kitchener, A.C. 1999. *The Distribution and Status of the Polecat* Mustela putorius *in Britain in the 1990s*. Vincent Wildlife Trust.

Briggs, B. & King, D. 1998. *The Bat Detective: A field guide to bat detection*. Batbox Ltd (CD & book).

Bright, P. 1991. *Where to Watch Mammals in Britain*. The Mammal Society.

Buczacki, S. 2002. *Fauna Britannica*. Hamlyn.

Carwardine, M. 2003. *Guide to Whale Watching: Britain and Europe*. Whale & Dolphin Conservation Society.

Churchfield, S. 1990. *The Natural History of Shrews*. Christopher Helm.

Clark, M. 2001. *Mammals, Amphibians and Reptiles of Hertfordshire*. Hertfordshire Natural History Society.

Corbet, G.B. 1989. *Finding and Identifying Mammals in Britain*. British Museum Natural History.

Corbet, G.B & Harris, S. 1991. *The Handbook of British Mammals*. 3rd edition. Blackwell Scientific Publications.

Corbet, G. & Ovenden, G. 1980. *The Mammals of Britain & Europe*. Collins.

Delaney, M.J. 1985. *Yorkshire Mammals*. Yorkshire Naturalists' Union.

Dobson, J. 1999. *The Mammals of Essex*. Lopinga Books.

Duff, A. & Lawson, A. 2004. *Mammals of the World: A Checklist*. A & C Black.

Dunstone, N. 1993. *The Mink*. T & AD Poyser.

Flowerdew, J. 1993. *Mice and Voles*. Whittet Books.

Gorman, M.L. & Stone, D 1990. *The Natural History of Moles*. Christopher Helm.

Harrap, S. & Redman, N. 2003. *Where to Watch Birds in Britain*. Christopher Helm.

Harris, S. & White, P. 1994. *The Red Fox*. The Mammal Society.

Harris, S. & McLaren, G. 1998. *The Brown Hare in Britain*. University of Bristol.

Harris, S., Morris, P., Wray, S. & Yalden, D. 1995. *A Review of British Mammals: Population estimates and conservation status of British mammals other than cetaceans*. JNCC.

Harrison Matthews, L. 1989. *British Mammals*. Collins New Naturalist Series.

Jones, K. & Walsh, A. 2001. *A Guide to British Bats*. Field Studies Council AIDGAP series.

King, C.M. 1989. *The Natural History of Weasels and Stoats*. Christopher Helm.

Kitchener, A. 1995. *The Wildcat*. The Mammal Society.

Lawrence, M.J. & Brown, R.W. 1974. *Mammals of Britain: Their tracks, trails and signs*. Blandford Press.

Macdonald, D.W. & Tattersall, F. 2001. *Britain's Mammals: The challenge for conservation*. Mammals Trust UK.

Macdonald, D.W. & Tattersall, F. 2002. *The State of Britain's Mammals 2002*. Mammals Trust UK.
Macdonald, D.W. & Tattersall, F. 2003. *The State of Britain's Mammals 2003*. Mammals Trust UK.
Macdonald, D.W. & Tattersall, F. 2004. *The State of Britain's Mammals 2004*. Mammals Trust UK.
Macdonald, D.W. & Baker, S. 2005. *The State of Britain's Mammals 2005*. Mammals Trust UK.
McDonald, R. & Harris, S. 1998. *Stoats & Weasels*. The Mammal Society.
Mitchell-Jones, A.J. *et al*. 1999. *The Atlas of European Mammals*. T & AD Poyser.
Morris, P. 1997. *The Edible Dormouse*. The Mammal Society.
Neal, E. & Cheeseman, C. 1996. *Badgers*. T & AD Poyser.
Reeve, N.J. 1994. *Hedgehogs*. T & AD Poyser.
Reeves, R.R., Stewart, B.S., Clapham, P.J. & Powell, J.A. 2002. *Sea Mammals of the World*. A&C Black.
Richardson, P. 2000. *Distribution Atlas of Bats in Britain and Ireland 1980–1999*. Bat Conservation Trust.
Sargent, G. & Morris, P. 2003. *How to Find and Identify Mammals*. The Mammal Society.
Schober, W. & Grimmberger, E. 1997. *The Bats of Europe and North America: knowing them, identifying them, protecting them*. T.F.H. Publications.
Shirihai, H. & Jarrett, B. 2006. *Whales, Dolphins and Seals: A Field Guide to the Marine Mammals of the World*. A&C Black.
Sterry, P. 1995. *Regional Wildlife: New Forest*. Dial House.
Thorburn, A. & Bishop, I. 1974. *Thorburn's Mammals*. Ebury Press.
Yalden, D. 1999. *The History of British Mammals*. T & AD Poyser.

Site Index

Species Index

Numbers in italics refer to the Species Accounts. Numbers in roman refer to the Site Accounts.